Table of conten'

INTRODUCTION

Instant Pot is one of the most innovative and multi-function kitchen appliances of today. It can replace usual pile of different cooking devices like: slow cooker, sauté-pan, oven, stove, steamer and much more. Let's point the main benefits it brings to your kitchen and everyday life:

- **Time saving** – pressure cooking method allows you to cook much faster than any other usual method saving your time for other activities or relaxation.
- **Energy efficiency** - not even saying about that the Instant Pot uses a lesser amount of water than any other appliances what consequently means less energy consumption, the fact that you will cook with a single appliance instead of multiple stoves and burners therefore saving a huge amount of energy.
- **Healthy cooking** – pressure cooking method also preserves wholesome nutrients and minerals at the same time eliminating many harmful micro-organisms. As the less water used through the cooking process the more vitamins and nutrients are not washed out with a water. Also the temperature maintained during pressure cooking is above the boiling point so most of the harmful bacteria and micro-organisms would be killed.
- **Programmable and versatile**– Instant Pot features a big variety of different cooking modes and regimes which allows to replace many ordinary appliances and also granting the ability to program your device up to 24 hours with delayed cooking.
- **Easy to wash and clean** – the cooking bowl inside the Instant Pot is made with a stainless steel providing reliability and easy cleaning.

In this book you will find everything you need to prepare the tastiest and delicious dishes you may wish to cook with your Instant Pot. I hope my book will bring you a lot of fun and will help to save your time and money.

My dear reader! In order to thank you for supporting me and my book I have prepared a free bonus for you (see the last page).

Scottish eggs

Deep into the atmosphere of Scotland while having breakfast! Nourishing eggs will award you with energy for the whole day.

Prep time: 5 minutes
Cooking time: 20 minutes
Servings: 3

Ingredients:

- 7 eggs
- 2/3 pound beef
- 2/3 pound pork
- 1 teaspoon salt

Directions:

1. Peel eggs. Take 2/3 pound beef and 2/3 pound pork and make a mince.
2. Add 1 egg to the mince and knead it.
3. Add 1 teaspoon salt (or any other spices if you want) and knead it once more.
4. Divide the mince into 6 equal pieces.
5. Then you have to turn each piece into flat cake in which we wrap an egg and seal it with a fork.
6. Place the flat cakes into the Instant Pot.
7. Press the Power Button. Choose program Meat/Stew.
8. The cooking time is 20 minutes.
9. Remove the Scottish eggs from Instant Pot when they are ready. Let it cool down.
10. Serve with any topping you want.

Nutrition:

- Calories: 253
- Fat: 61g
- Carbohydrates: 32g
- Protein: 44g

Camambert with ham

The gentle taste of camembert with nourishing ham will make your morning unforgettable. Delicious breakfast for all family

Prep time: 20 minutes
Cooking time: 45 minutes
Servings: 8

Ingredients:

- 2 pound camambert
- 1/3 pound ham
- 1 egg
- 1 cup flour
- 1/3 pound breadcrumbs
- 4 ounce confiture
- 2 cup orange juice
- 1 teaspoon mustard

Directions:

1. Firstly, you need to beat an egg. Place flour and breadcrumbs into separate plates.
2. The ham must be cut into 4 equal pieces. Cut a block of cheese in half and put the ham between the halves. Press it lightly.
3. Dredge the cheese and ham with flour. Then dip into egg and dredge with breadcrumbs.
4. Put the cheese into fridge for 20 minutes.
5. While the cheese is in the fridge, you should make a sauce. Mix confiture, juice and mustard in a bowl.
6. After that, get the cheese out of fridge and place it into Instant Pot.
7. Press the Power Button and arrange the needed time. The time is 45 minutes.
8. Choose program Meat/Stew.
9. Then remove camembert from Instant Pot and place it on a napkin in order to absorb a fat.
10. Serve with ketchup or mustard.

Nutrition:

- Calories: 272
- Fat: 18g
- Carbohydrates: 13g
- Protein: 16g

Mozzarella with vegetables

The piece of Italy on your table! Enjoy the tasty meal with a source of vitamins and minerals. This dish will satisfy the most demanding gourmet

Prep time: 50 minutes
Cooking time: 30 minutes
Servings: 8

Ingredients:

- 1 pound mozzarella
- 1 cucumber
- 2 tomatoes
- 1 onion
- 1 ounce pepper
- 1 cup flour
- 1 egg
- 1/3 pound breadcrumbs

Directions:

1. Peel tomatoes and cut it into 4 pieces.
2. Slice cucumber.
3. Mix tomatoes and courgettes in a bowl and put it aside.
4. Mix flour with pepper in a bowl. Beat the egg in a separate container.
5. Cut cheese into cubes and dredge them in flour. Then dip it into egg and dredge with breadcrumbs.
6. Place cheese into Instant Pot. Press the Power Button. Choose program Meat/Stew
7. Set the time – 30 minutes.
8. When mozzarella is ready, remove it from Instant Pot.
9. Serve cheese with a garnish of vegetables.

Nutrition:

- Calories: 300
- Fat: 24g
- Carbohydrates: 10g
- Protein: 9g

Shelpek

This dish of Kazakh cuisine will inspire you for the whole day. A kind of pancakes in the morning is a key to successful day.

Prep time: 10 minutes
Cooking time: 35 minutes
Servings: 8

Ingredients:

- ¾ cup water
- 2 cup flour
- 1 teaspoon salt

Directions:

1. Mix flour, salt and water in a bowl. Knead the dough. The dough should be elastic and soft.
2. Put it aside for 10 minutes.
3. Then divide the dough into 7 parts. Roll out each part of dough. The thickness should be approximately 3/64 inch.
4. Make from a dough small flat cakes.
5. Place the flat cakes into Instant Pot. Press the Power Button. Choose program Meat/Stew.
6. Set the time which is 35 minutes.
7. Remove the flat cakes from Instant Pot. Shelpek is ready.
8. Serve with any topping including honey, chocolate or jam.

Nutrition:

- Calories: 378
- Fat: 23g
- Carbohydrates: 37g
- Protein: 7g

Fried Suluguni Cheese Sticks

An ideal breakfast for 2 persons. Crispy suluguni cheese will create a lovely atmosphere in your home and bring lots of positive emotions.

Prep time: 5 minutes
Cooking time: 45 minutes
Servings: 2

Ingredients:

- 8 ounce suluguni cheese
- 1 egg
- 1.5 tablespoon cream
- ¾ cup flour

Directions:

1. Cut cheese into strips. The thickness is approximately 0.3 inches, and the length is 2 inches.
2. Beat the egg with cream.
3. Dredge cheese with flour. Then dip it into egg. After that, dredge cheese with flour ounce more.
4. Repeat the third step with all pieces of cheese.
5. Place the cheese into Instant Pot. Press the Power Button. Choose program Meat/Stew.
6. Set the needed time which is 45 minutes.
7. Remove cheese sticks from Instant Pot and let it cool down.
8. Serve with ketchup or sour cream.

Nutrition:

- Calories: 278
- Fat: 13g
- Carbohydrates: 37g
- Protein: 7g

Fried Champignons

Champignons for breakfast? Why not?! The true source of protein is waiting for you. Not so many efforts to taste this dish. Try it!

Prep time: 5 minutes
Cooking time: 45 minutes
Servings: 2

Ingredients:

- 1 pound champignons
- 2 egg
- 1 cup flour
- ½ cup milk
- 4 ounce breadcrumbs
- 1 teaspoon pepper
- 1 teaspoon salt

Directions:

1. Peel champignons.
2. Beat the eggs with milk. Add pepper and salt.
3. Dip champignons into egg. Then dredge it with flour. After that dip champignons into egg once more.
4. Dip champignons into breadcrumbs. After that, dip into egg one more time.
5. Place champignons into Instant Pot. Press the Power Button. Choose program Meat/Stew.
6. Arrange the needed time. Time – 45 minutes.
7. Remove champignons from Instant Pot when they are ready. Put them on napkin in order to absorb the fat.
8. Serve with any topping or garnish.

Nutrition:

- Calories: 175
- Fat: 11g
- Carbohydrates: 17g
- Protein: 10g

Baursaks with stuffing

So sweet morning cannot do without sweet breakfast! Pamper yourself with nice Asian donuts with apricot.

Prep time: 1.5 hour
Cooking time: 40 minutes
Servings: 4

Ingredients:

- 4.5 cup flour
- 1 cup sugar
- 1/3 pound butter
- 1 ounce yeast
- 2 ounce almonds
- 1 cup milk
- 2 eggs
- 2 tablespoon rom
- 0.5 teaspoon cinnamon
- 30 apricots

Directions:

1. Dissolve sugar and yeast in warm milk. Put it aside for 10 minutes.
2. Then add 2 ounce butter, flour, eggs, 1 tablespoon rom and knead a dough. Put it aside for 1.5 hours. It should double its size.
3. While a dough is "resting" make a stuffing. Grind almonds. Add 2 ounce butter, 2 ounce sugar, 1 tablespoon flour, 2 teaspoon rom and 1 cup milk. Stir it well. Place it in fridge for 30 minutes.
4. Extract apricot kernels. Put the stuffing from almond into apricot instead of kernels. Fasten two halves of apricot.
5. Roll out a dough and cut out small circles. Put stuffing onto each circle. Seal the edges with a fork. It looks like round flatbread.
6. Place these flatbreads into Instant Pot and press the Power Button. Choose program Meat/Stew.
7. Set a time – approximately 40 minutes.
8. Remove flatbreads from Instant Pot and let it cool down.
9. Tasty Asian donuts are ready. Also, you can spread cinnamon.

Nutrition:

- Calories: 475
- Fat: 51g
- Carbohydrates: 67g
- Protein: 25g

Mini sausages in dough

Like hot-dogs only smaller! Delicious sausages in dough are perfect variant for breakfast, aren't they?

Prep time: 5 minutes
Cooking time: 45 minutes
Servings: 2

Ingredients:

- 1 pound filo pastry
- 5 sausages

Directions:

1. Defrost a filo pastry.
2. Roll out a filo pastry. Divide it into 4 parts.
3. Cut each part of filo pastry into halves to make small squares.
4. Dice sausages and put them onto pieces of filo pastry. Clench the edges.
5. Slice each square of filo pastry with sausage.
6. Place sausages into Instant Pot. Pres the Power Button. Choose program Meat/Stew.
7. Set a time which is 45 minutes.
8. Remove the sausages from Instant Pot and let them cool down.
9. Serve with ketchup or mustard.

Nutrition:

- Calories: 235
- Fat: 15g
- Carbohydrates: 25g
- Protein: 8g

Curd donuts

It is a perfect variation of breakfast with curd. Gentle and delicious donuts could be the most memorable part of morning.

Prep time: 5 minutes
Cooking time: 40 minutes
Servings: 2

Ingredients:

- 8 ounce curd
- 1 cup flour
- 1 egg
- 5 tablespoon sugar

Directions:

1. Whip curd in order to make it smooth.
2. Add sugar.
3. Crack egg into curd. Stir it thoroughly.
4. Then sift a flour and add to curd. Knead a soft dough.
5. Roll the tourniquet from dough. Then divide it into small pieces and roll balls.
6. Place balls into Instant Pot and press the Power Button. Choose program Meat/Stew.
7. Arrange the needed time: time – 40 minutes.
8. Remove balls from Instant Pot and let it cool down.
9. Curd donuts are ready. Serve them with sour cream or jam.

Nutrition:

- Calories: 185
- Fat: 12g
- Carbohydrates: 18g
- Protein: 9g

Cherry patty

Sunday morning…What could be better? Only Sunday morning with cherry petty! Delicious and crispy patty with marvelous cherries are designed for ideal moments.

Prep time: 30 minutes
Cooking time: 25 minutes
Servings: 2

Ingredients:

- 1 cup flour
- 1 pound cherries
- 4 ounce butter
- ¼ cup water
- ½ teaspoon salt
- 1 tablespoon farina
- 3 tablespoon milk

Directions:

1. Cut butter into small bits. Add it to flour.
2. Then grind butter and flour into crumb. The dough should be very crumbly.
3. Add cold water. Knead an elastic dough. Put it into fridge for 30 minutes.
4. Make a stuffing. Cut cherries if they are too big. Mix cherry with sugar and water. Put it on a cooker. Make this mass thick.
5. Roll out a dough. Cut it into long rectangles. Lay the stuffing on the center of each rectangle.
6. Seal each rectangle with a fork.
7. Put them into Instant Pot. Press the Power Button. Choose program Meat/Stew.
8. Set a time which is 45 minutes.
9. Remove patties from Instant Pot when they are ready. Let it cool down.
10. You can serve cherry petties with chocolate or honey.

Nutrition:

- Calories: 265
- Fat: 18g
- Carbohydrates: 27g
- Protein: 10g

Patty cakes with plums

Plums are reach source of vitamins. And patty cakes are rich source of happiness. It is a perfect mixture for breakfast, isn't it?

Prep time: 5 minutes
Cooking time: 35 minutes
Servings: 2

Ingredients:

- 8 ounce filo pastry
- 8 ounce plums
- 1 cup sugar

Directions:

1. Cut plums.
2. Cut filo pastry into long rectangles.
3. Put plums onto center of each rectangle. Seal them with fork.
4. Spread rectangles with sugar.
5. Place patty cakes into Instant Pot Press the Power Button. Choose program Sauté/browning.
6. Arrange the temperature which 330 degrees and set a time – 35 minutes.
7. Remove patty cakes when they are ready. Let them cool down.
8. You can serve patty cakes with any jam or chocolate. It depends on your taste.

Nutrition:

- Calories: 245
- Fat: 15g
- Carbohydrates: 22g
- Protein: 7g

Samos

This delicious dish is an implementation of harmony of tastes! Yummy cakes with vegetables – your choice of useful breakfast.

Prep time: 20 minutes
Cooking time: 35 minutes
Servings: 2

Ingredients:

- 5 cup flour
- 5.5 ounce butter
- 1 cup water
- 4 potatoes
- 1 cauliflower
- 8 ounce green peas
- 2 teaspoon chili pepper

Directions:

1. Sift a flour. Add salt and butter.
2. Add the melted butter. Add salt.
3. Rub the mixture with your fingers until crumbs are obtained.
4. Add hot water and knead the dough. It should be soft and elastic.
5. Roll the dough into a ball. Drizzle it with water and cover with a damp clingfilm. Put it aside for 20 minutes.
6. Meanwhile, make stuffing. Peel and chop potatoes.
7. Slice all vegetables.
8. Mix all vegetables and stir it well.
9. Roll out the dough and cut it into balls.
10. Roll out these balls and cut them in half.
11. Roll the rods and clasp them from the bottom.
12. Fill the rods with stuffing (mixture of vegetables) and seal them with a fork.
13. Put the rods into Instant Pot and press the Power Button. Choose program Sauté/browning.
14. Set a time at 35 minutes.
15. Remove rods from Instant Pot when the time is over. Let them cool down.
16. Samos are ready. It is the best way to serve them with ketchup.

Nutrition:

- Calories: 180
- Fat: 13g
- Carbohydrates: 17g
- Protein: 8g

Fried cauliflower

If you are fond of useful and diet breakfast, then fried cauliflower is the best variant you've ever chosen. Owing to its easiness in cooking, it could become a real rescuer in the morning.

Prep time: 30 minutes
Cooking time: 35 minutes
Servings: 3

Ingredients:

- 1 curd cauliflower
- 2 yolks
- 1/2 cup milk
- 2.5 cup flour
- 1 teaspoon white pepper
- 1 teaspoon salt

Directions:

1. Beat the yolks. Add salt and pepper.
2. Add 4 ounce flour gradually. Knead batter.
3. Add milk to batter and knead it once more.
4. Put batter into fridge for 30 minutes.
5. After that, put cauliflower into batter.
6. Place cauliflower into Instant Pot. Press the Power Button. Choose program Sauté/browning.
7. Set a time – 15 minutes and arrange the temperature – 330 degrees.
8. Remove cauliflower from Instant Pot. Let it cool down.
9. Fried cauliflower is ready.

Nutrition:

- Calories: 130
- Fat: 9g
- Carbohydrates: 12g
- Protein: 5g

Fried corn in dough

The delicious taste and smell of this dish will make anybody wake up.
Check it by yourself!

Prep time: 10 minutes
Cooking time: 40 minutes
Servings: 2

Ingredients:

- 1 corn cob
- 1 cup flour
- 1 egg
- 1 ounce butter
- ½ cup milk
- 1 teaspoon salt

Directions:

1. Separate the white from the yolk.
2. Grind yolks with sugar and salt. Add milk.
3. Sift a flour and add to yolks with milk.
4. Add whipped whites and knead a dough.
5. Put it aside for 10 minutes.
6. Meanwhile, peel corn and dice it.
7. Roll out a dough and cut circles.
8. Put corn into the center of each circle. Seal with a fork.
9. Place circles into Instant Pot. Press the Power Button. Choose program Sauté/browning.
10. Set a time which is 40 minutes.
11. When the time is over, remove the corn from Instant Pot. Let it cool down.
12. Fried corn in dough is ready. Enjoy it.

Nutrition:

- Calories: 170
- Fat: 13g
- Carbohydrates: 15g
- Protein: 8g

Apples in batter

If Snow White ate these apples she wouldn't fall asleep! Pamper yourself with delicious breakfast made of apples in batter

Prep time: 10 minutes
Cooking time: 25 minutes
Servings: 2

Ingredients:

- 1 pound apples
- ½ lemon
- 2 tablespoon rom
- 12 ounce sugar

Directions:

1. Peel and cut apples. Put it into bowl.
2. Drizzle it with lemon.
3. Add rom and sugar.
4. Cover bowl with another plate and put it aside for 10 minutes.
5. Place apples into Instant Pot. Press the Power Button. Choose program Sauté/browning
6. Arrange the settings: time – 25 minutes.
7. After that, remove apples from Instant Pot. Let them cool down.
8. If you are a permanent sweet-tooth, you can sprinkle apples with sugar.

Nutrition:

- Calories: 120
- Fat: 10g
- Carbohydrates: 15g
- Protein: 6g

Pineapple balls

Prolong your weekends and make week days the sweetest in the world! Crispy pineapple balls will help you to achieve this goal.

Prep time: 5 minutes
Cooking time: 30 minutes
Servings: 4

Ingredients:

- 1 egg
- 4 ounce flour
- ¼ cup milk
- 1 pineapple
- 4 ounce sugar
- 1 teaspoon salt

Directions:

1. Beat egg and add sugar.
2. Add sifted flour and milk
3. Knead a dough.
4. Grate pineapple and add to dough.
5. Shape balls.
6. Place balls into Instant Pot. Press the Power Button. Choose program Sauté/browning
7. Set a needed time which is 30 minutes.
8. When the time is over, remove pineapple balls from Instant Pot.
9. Let them cool down.
10. You can serve pineapple balls with chocolate or honey to make them sweeter.

Nutrition:

- Calories: 160
- Fat: 12g
- Carbohydrates: 19g
- Protein: 9g

<u>Eggs with ham</u>

Traditional breakfast with untraditional cooking! This transformation will surprise anybody by its delicious and nutritious taste.

Prep time: 5 minutes
Cooking time: 30 minutes
Servings: 2

Ingredients:
- 1 pound ham
- 4 eggs
- 1 teaspoon flour
- 4 ounce breadcrumbs

Directions:
1. Chop ham.
2. Beat eggs.
3. Mix flour and breadcrumbs.
4. Add ham and eggs to mixture of flour and breadcrumbs.
5. Make medium doughboys from this mass.
6. Place these doughboys into Instant Pot Press the Power Button. Choose program Meat/Stew.
7. Arrange the settings: time – 30 minutes.
8. Ehen the time is over, remove doughboys from Instant Pot.
9. Let them cool down.
10. Serve them with ketchup or sour cream. Bon appetit!

Nutrition:
- Calories: 193
- Fat: 15g
- Carbohydrates: 25g
- Protein: 11g

<u>German eggs with salad</u>

Guten morgen alle! Try these delicious German breakfast and deep into its culture for 15 minutes!

Prep time: 5 minutes
Cooking time: 45 minutes
Servings: 4

Ingredients:

- 1 white
- 4 eggs
- 2 tablespoon flour
- 3 tablespoon breadcrumbs
- 1 teaspoon pepper
- Green salad
- Lettuce
- 1 chili
- 2 tomatoes

Directions:

1. Peel eggs. Dredge them with flour.
2. Then, dredge them with white.
3. Put eggs into breadcrumbs.
4. Place eggs into Instant Pot. Press the Power Button. Choose program Meat/Stew.
5. Set a time at 45 minutes.
6. Remove eggs from Instant Pot. Let them cool down.
7. Rinse both sorts of salad and cut them.
8. Peel and cut tomatoes.
9. Mix tomatoes and salad. Add pepper and salt.
10. Serve eggs with this salad. Enjoy the meal!

Nutrition:

- Calories: 230
- Fat: 15g
- Carbohydrates: 27g
- Protein: 13g

Eggs in dough

Eggs + dough = nutritious breakfast. It is a perfect formula, isn't it?

Prep time: 5 minutes
Cooking time: 40 minutes
Servings: 4

Ingredients:

- 10 eggs
- ½ cup flour
- 1/3 cup milk
- 1 teaspoon salt
- 1 ounce butter
- 1 teaspoon pepper

Directions:

1. Separate two yolks from whites.
2. Add salt, butter and yolks to flour. Stir it well.
3. Whip the whites and add to flour. Add pepper. Mix it well.
4. Peel boiled eggs and dip into dough.
5. Place eggs into Instant Pot and press the Power Button. Choose program Meat/Stew.
6. Set a time – 40 minutes.
7. When the time is over, remove eggs from Instant Pot.
8. Eggs in dough are ready. You can serve them with any vegetables.

Nutrition:

- Calories: 240
- Fat: 16g
- Carbohydrates: 26g
- Protein: 14g

Sausages from cottage cheese

Cottage cheese is an ideal product for breakfast. In the form of sausages, it will become more delicious and unusual.

Prep time: 5 minutes
Cooking time: 40 minutes
Servings: 2

Ingredients:

- 1 pound cottage cheese
- 1/3 pound oat flour
- 4 ounce honey
- 2 eggs
- 1 ounce flour
- 4 ounce sour cream

Directions:

1. Mix cottage cheese with eggs, oat flour and honey.
2. Roll out this mass in the form of sausage.
3. Cut sausage into pieces and dredge each of them with flour.
4. Place sausages into Instant Pot and press the Power Button. Choose program Meat/Stew.
5. Set a time – 40 minutes.
6. Remove sausages, when the time is over. Let them cool down.
7. Serve them with sour cream.

Nutrition:

- Calories: 220
- Fat: 13g
- Carbohydrates: 22g
- Protein: 10g

Cottage cheese bars

Not only chocolate bars could be eaten with pleasure – cottage cheese bars are an ideal breakfast and useful substitute of chocolate bars.

Prep time: 5 minutes
Cooking time: 40 minutes
Servings: 2

Ingredients:

- 4 ounce cottage cheese
- 2 ounce flour
- 1 egg
- 1 teaspoon sugar
- 1 tablespoon sour cream
- 1 teaspoon salt

Directions:

1. Add flour and egg into wiped cottage cheese.
2. Stir it well.
3. Add sour cream and sugar. Stir it well.
4. Roll out this mass. The thickness should be approximately 0.2 inches. Cut it into strips.
5. Place strips into Instant Pot. Press the Power Button. Choose program Meat/Stew.
6. Arrange the settings: time – 40 minutes.
7. When the time is over, remove strips from Instant Pot.
8. Cottage cheese bars are ready. You can serve them with honey or jam. Have a nice meal!

Nutrition:

- Calories: 180
- Fat: 12g
- Carbohydrates: 16g
- Protein: 8g

Cottage cheese balls

Gentle and sweet cottage cheese balls have a wonderful power – they could improve your mood in the early morning!

Prep time: 5 minutes
Cooking time: 45 minutes
Servings: 1

Ingredients:

- 8 ounce cottage cheese
- 1 cup flour
- 4 eggs
- 3 tablespoon sugar
- 3 tablespoon sour cream

Directions:

1. Beat eggs. Add sugar and flour. Beat once more.
2. Add cottage cheese and sour cream.
3. Knead an elastic dough.
4. Roll out a dough and shape small circles.
5. Place circles into Instant Pot. Press the Power Button. Choose program Meat/Stew.
6. Set a time which is 45 minutes.
7. Remove balls, when the time is over.
8. Let them cool down and serve with any topping!

Nutrition:

- Calories: 170
- Fat: 10g
- Carbohydrates: 15g
- Protein: 6g

Pancakes

Gentle and sweet cottage cheese balls have a wonderful power – they could improve your mood in the early morning!

Prep time: 5 minutes
Cooking time: 30 minutes
Servings: 2

Ingredients:

- 1 pound potatoes
- 8 ounce cottage cheese
- 1 cup flour
- 3 ounce sugar
- 1 egg
- Lemon zest

Directions:

1. Peel potatoes and make mashed potatoes.
2. Add flour and egg. Stir it well.
3. Add cottage cheese and sugar. Knead and elastic dough.
4. Roll out a dough and cut medium of size circles.
5. Place circles into Instant Pot and press the Power Button. Choose program Sauté/browning.
6. Set a time which is 30 minutes.
7. Remove pancakes when the time is over.
8. Serve with jam or sour cream. Enjoy your meal!

Nutrition:

- Calories: 210
- Fat: 15g
- Carbohydrates: 23g
- Protein: 8g

Croquettes from cream cheese

The gentleness of cream cheese with crispiness of croquettes is a guarantee of a perfect breakfast.

Prep time: 5 minutes
Cooking time: 20 minutes
Servings: 2

Ingredients:

- 1/3 pound cream cheese
- 3 tablespoon margarine
- 2 ounce bread
- 3 tablespoon farina
- 1 yolk
- 1/3 pound breadcrumbs

Directions:

1. Whip margarine.
2. Mix margarine with cream cheese.
3. Add soaked and squeezed bread.
4. Add yolk and farina.
5. Then you have to divide this mass into pieces. Dredge them with breadcrumbs.
6. Make pieces longer like thin strips.
7. Place croquettes into Instant Pot. Press the Power Button. Choose program Meat/Stew.
8. Set e time – 20 minutes.
9. Remove croquettes when time is over.
10. Croquettes from cream cheese are ready. Bon appetit!

Nutrition:

- Calories: 230
- Fat: 17g
- Carbohydrates: 24g
- Protein: 10g

French cheese balls

Notes of France on your table! Try this delicious dish of French cuisine and feel yourself as gourmet.

Prep time: 5 minutes
Cooking time: 30 minutes
Servings: 1

Ingredients:

- 4 ounce cheese
- 4 oucne parsley
- 1 teaspoon salt
- 1 teaspoon pepper
- ½ cup flour

Directions:

1. Grate cheese. Put it into bowl.
2. Add salt and pepper to cheese.
3. Shape balls from this mass.
4. Dredge balls with flour.
5. Place cheese balls into Instant Pot. Press the Power Button. Choose program Meat/Stew.
6. Arrange the settings: time – 30 minutes.
7. When the time is over, remove cheese balls.
8. Cheese balls should be hot. Serve them with sour cream and parsley.

Nutrition:

- Calories: 210
- Fat: 15g
- Carbohydrates: 22g
- Protein: 9g

Delisez from cheese

If you are not in France, still there is a possibility to wake up in its atmosphere! Cook this French dish for a breakfast and deep into atmosphere of Louvre and the Eiffel Tower

Prep time: 5 minutes
Cooking time: 45 minutes
Servings: 1

Ingredients:

- 4 whites
- 8 oucne cheese
- 1/3 pound breadcrumbs
- 4 ounce parsley
- 1 teaspoon pepper

Directions:

1. Whip whites.
2. Grate cheese.
3. Mix cheese and whites.
4. Add pepper.
5. Shape balls from this mass.
6. Each ball dredge with breadcrumbs.
7. Place them into Instant Pot. Press the Power Button. Choose program Meat/Stew.
8. Set a time which is 45 minutes.
9. Remove balls when the time is over.
10. Let them cool down.
11. Delisez from cheese is ready. Serve with parsley and sour cream.

Nutrition:

- Calories: 190
- Fat: 14g
- Carbohydrates: 21g
- Protein: 8g

Pumpkin wafers

Even if it is quite far to Halloween, pumpkin is a great product for tasty dishes. Especially, for breakfast. Owing to its richness in vitamin, it is a perfect variant of meal in the morning.

Prep time: 30 minutes
Cooking time: 10 minutes
Servings: 4

Ingredients:

- 1 pumpkin
- 1 teaspoon salt
- 1 teaspoon turmeric
- 1/4 teaspoon ground hot red pepper
- 3 tablespoon rice flour

Directions:

1. Peel pumpkin and cut it into small pieces. Put it into bowl.
2. Add salt and pepper. Put it aside for 30 minutes.
3. After that, dredge pieces with rice flour.
4. Place pieces of pumpkin into Instant Pot. Press the Power Button. Choose program Sauté/browning.
5. Set a time – 10 minutes.
6. When time is over, remove pumpkin and let it cool down.
7. Pumpkin wafers are ready. Have a nice meal!

Nutrition:

- Calories: 210
- Fat: 12g
- Carbohydrates: 17g
- Protein: 8g

Zucchini with peanuts

Even if it is quite far to Halloween, pumpkin is a great product for tasty dishes. Especially, for breakfast. Owing to its richness in vitamin, it is a perfect variant of meal in the morning.

Prep time: 30 minutes
Cooking time: 40 minutes
Servings: 2

Ingredients:

- 1 cup pea flour
- 1 tablespoon lemon juice
- 1 teaspoon seeds of Indian caraway seeds
- 1 teaspoon hot red pepper
- 1 teaspoon salt
- 2/3 cup water
- 3 tablespoon chopped peanuts
- ¾ pound zucchini

Directions:

1. Cut zucchini
2. Mix pea flour, lemon juice, seeds of Indian caraway seeds and pepper. Stir it well.
3. Pour water gradually. Whip it slightly. It should be very smooth dough. Put aside for 10 minutes. And whip it once more.
4. Add peanuts.
5. Dip the pieces of zucchini into dough.
6. Place zucchini into Instant Pot. Press the Power Button. Choose program Sauté/browning.
7. Set a time – 40 minutes.
8. Remove zucchini from Instant Pot.
9. Zucchini with peanuts are ready. Bon appetite!

Nutrition:

- Calories: 160
- Fat: 11g
- Carbohydrates: 15g
- Protein: 6g

Dutch Benete

Dutch national dish is devoted to be on your table for a breakfast! Easy to cook and easy to eat, it could become the best breakfast you've ever tried.

Prep time: 30 minutes
Cooking time: 45 minutes
Servings: 4

Ingredients:

- 2 cup flour
- 1 teaspoon salt
- ½ cup milk
- 0.5 ounce brewer's yeast,
- 1 ounce butter
- 1 ounce sugar
- 1 ounce raisins
- 1 grated apple
- 2 ounce lemon zest
- 1 egg

Directions:

1. Dissolve the brewer's yeast in warm milk.
2. Mix it with flour, salt, egg, butter and sugar. Knead a dough.
3. Add grated apple and raisins. Put it aside for 1 hour.
4. After that, roll out a dough and shape small balls.
5. Place balls into Instant Pot. Press the Power Button. Choose program Sauté/browning.
6. Seta time at 45 minutes.
7. When the time is over, remove balls from Instant Pot. Let them cool down.
8. Dutch Benete is ready. You can sprinkle balls with sugar glaze.

Nutrition:

- Calories: 220
- Fat: 15g
- Carbohydrates: 23g
- Protein: 9g

Banana slices in beer batter

Banana and beer sounds strange, isn't it? Not for tasty breakfast! This dish will surprise you by harmony of flavors

Prep time: 5 minutes
Cooking time: 40 minutes
Servings: 2

Ingredients:

- 4 bananas
- 4 tablespoon lemon juice
- 1 egg
- 4 tablespoon sugar
- 1 cup flour
- ½ cup beer
- 1 ounce raisins
- 1 grated apple
- 2 ounce lemon zest
- 1 egg

Directions:

1. Peel bananas and cut them into 3 pieces.
2. Pour bananas with lemon juice.
3. Mix egg and sugar. Add flour and beer gradually. Whip it well. Put it aside for 10 minutes.
4. Put bananas into this mass.
5. Place bananas into Instant Pot and press the Power Button. Choose program Sauté/browning.
6. Arrange the needed time: time – 40 minutes.
7. When the time is over, remove bananas and let them cool down.
8. Serve with any topping. Also, you can sprinkle bananas with sugar.

Nutrition:

- Calories: 170
- Fat: 14g
- Carbohydrates: 19g
- Protein: 7g

Malaysian potato donuts

Everything as you like: crispy gentle donuts and soft potato. Don't hesitate to cook the dish of Malaysian cuisine.

Prep time: 5 minutes
Cooking time: 45 minutes
Servings: 4

Ingredients:

- 1.5 pound potatoes
- 1 cup flour
- 3 tablespoon water
- 4 tablespoon sesame

Directions:

1. Make puree from potatoes.
2. Add flour, sugar and sesame.
3. Stir it well, grinding the flour with potatoes.
4. Add water gradually. The dough should be elastic.
5. Roll out the dough and shape small balls/donuts.
6. Place donuts into Instant Pot. Press the Power Button. Choose program Sauté/browning.
7. Set time – 45 minutes.
8. Remove donuts and let them cool down.
9. You can serve them with ketchup or mustard.

Nutrition:

- Calories: 123
- Fat: 10g
- Carbohydrates: 17g
- Protein: 7g

Milk porridge with raisins

Porridges are very beneficial kind of breakfast. This one isn't only nutritious but also a tasty one.

Prep time: 5 minutes
Cooking time: 40 minutes
Servings: 4

Ingredients:

- 2 cup oat flakes
- 5 cup milk
- 1 cup raisins
- 1 tablespoon salt
- 2 tablespoon sugar

Directions:

1. Mix oatflakes, sugar and salt.
2. Add milk.
3. Add raisins. Stir it well.
4. Put it into Instant Pot. Press the Power Button.
5. Choose Porrdige program. Set a time at 40 minutes.
6. When the time is over, remove porridge from Instant Pot.
7. Milk porridge with raisins is ready. Mission is done.

Nutrition:

- Calories: 123
- Fat: 10g
- Carbohydrates: 17g
- Protein: 7g

Pearl barley porridge with butter

Many people do not like pearl porridge, but they should! This porridge is very useful. It will not be difficult to cook it at all.

Prep time: 4 hours
Cooking time: 1 hour
Servings: 4

Ingredients:

- 2 cup pearl barley
- 2 ounce butter
- 4.5 cup water
- 1 tablespoon salt

Directions:

1. Wash pearl barley. Pour with boiling water.
2. Put it aside for 4 hours.
3. After that, place pearl barley into Instant Pot.
4. Choose Porridge program and set a time at 1 hour.
5. Add butter.
6. When the time is over, remove pearl barley from Instant Pot.
7. Pearl barley with butter is ready. Bon appetite!

Nutrition:

- Calories: 133
- Fat: 11g
- Carbohydrates: 15g
- Protein: 7g

Millet gruel

Millet is the source of many vitamins that strengthen our body. So, it is a crime not to cook it for a breakfast.

Prep time: 4 hours
Cooking time: 1 hour
Servings: 4

Ingredients:

- 1 cup millet
- 3 cup milk or water
- 1 tablespoon salt
- 2 tablespoon sugar
- 2 tablespoon butter

Directions:

1. Mix millet and sugar.
2. Add water or milk,
3. Add salt.
4. Stir it well.
5. Place it into Instant Pot. Press the Power Button.
6. Choose Porridge program and set time at 30 minutes.
7. Add butter.
8. When the time is over, remove porridge from Instant Pot.
9. Millet gruel is ready. Have a nice meal!

Nutrition:

- Calories: 123
- Fat: 10g
- Carbohydrates: 17g
- Protein: 7g

Steam beef cutlets

It is a well-known fact that dished cooked in steam possess more vitamins and minerals than other dishes. That's why it is an obligatory to eat such dishes when you are on diet or just want to eat healthy food.

Prep time: 5 minutes
Cooking time: 45 minutes
Servings: 2

Ingredients:

- 2 ounce butter
- ½ cup milk
- 1 clove of garlic
- 1 onion
- 1 egg
- 4 ounce breadcrumbs
- 1 tablespoon salt and pepper

Directions:

1. Wash and grind beef. Make a mince.
2. Season with salt and pepper.
3. Add chopped onion to meat.
4. And milk and egg to mince. Stir it well.
5. Shape small cutlets from mince.
6. Dredge cutlets with breadcrumbs
7. Place them into Instant Pot. Press the Power Button.
8. Choose Steam program.
9. Set a time – 45 minutes.
10. When the time is over, remove cutlets from Instant Pot.
11. The dish is ready. Serve it with rice.

Nutrition:

- Calories: 105
- Fat: 8g
- Carbohydrates: 12g
- Protein: 6g

Porridge from bulgar

Bulgur is a very useful cereal. Porridge from bulgur is very easy to cook. So we do not see any obstacles. Let's go!

Prep time: 5 minutes
Cooking time: 55 minutes
Servings: 2

Ingredients:

- 1 ounce butter
- 1/3 pound bulgur
- 2 cup cold water
- 2 teaspoon sugar
- 1 teaspoon salt, pepper

Directions:

1. Place butter into Instant Pot.
2. Add bulgur.
3. Choose program Multigrain and set a time at 15 minutes.
4. Then add water.
5. Add sugar, salt and pepper.
6. Choose program Porridge.
7. Set time at 40 minutes.
8. When the time is over, remove porridge from Instant Pot.
9. Porridge from bulgar is ready. Enjoy your meal!

Nutrition:

- Calories: 135
- Fat: 10g
- Carbohydrates: 14g
- Protein: 7g

Semolina with carrot juice

Very interesting combination of semolina and juice. Pamper yourself with vitamins and minerals!

Prep time: 5 minutes
Cooking time: 55 minutes
Servings: 2

Ingredients:

- 1 carrot
- 2 tablespoon semolina
- 2/3 cup milk
- ¾ cup water
- 1 tablespoon salt,
- 2 tablespoon granulated sugar

Directions:

1. Mix semolina and milk. Stir it well.
2. Add water.
3. Place it into Instant Pot. Choose program Slow Cook.
4. Set a time at 15 minutes.
5. Add salt and sugar.
6. Choose program Porridge and set a time at 40 minutes
7. Grate carrot.
8. Add 1 tablespoon sugar to carrot.
9. When the time is over, remove porridge from Instant Pot.
10. Add carrot to semolina. The dish is ready. Bon appetite!

Nutrition:

- Calories: 115
- Fat: 10g
- Carbohydrates: 13g
- Protein: 7g

Oatmeal with apple and cinnamon

Oatmeal with an apple and cinnamon has an unusual and original taste. Porridge is useful, tasty, fragrant and very nourishing!

Prep time: 5 minutes
Cooking time: 35 minutes
Servings: 2

Ingredients:

- 1 cup oatmeal
- 1 cup milk
- 1 cup water
- 1 apple
- 1 ounce cinnamon
- 1 tablespoon salt
- 2 tablespoon granulated sugar

Directions:

1. Dice apple.
2. Mix apple and oatmeal.
3. Add milk.
4. Add cinnamon, salt and sugar.
5. Place it into Instant Pot. Choose Porridge program.
6. Set a time at 35 minutes.
7. After 20 minutes, add water.
8. When the time is over, remove porridge from Instant Pot.
9. Oatmeal with apple and cinnamon is ready. Bon appetite!

Nutrition:

- Calories: 88
- Fat: 7g
- Carbohydrates: 11g
- Protein: 5g

Rice-millet porridge

Rice, millet, pumpkin and apple - it is the perfect combination, isn't it? Well, it looks very attractive and the taste is not less impressive than its look. Try it!

Prep time: 5 minutes
Cooking time: 35 minutes
Servings: 2

Ingredients:

- ½ cup rice
- ½ cup millet
- 4 ounce pumpkin
- 1 apple
- ¾ cup milk
- 2/3 cup water
- 1 teaspoon sugar
- 1 teaspoon salt

Directions:

1. Mix rice and millet.
2. Peel and dice apples.
3. Add apples to a mixture of rice and millet.
4. Grind pumpkin and add to rice and millet.
5. Place it into Instant Pot. Choose Porridge program.
6. Set a time at 35 minutes.
7. Add milk and water.
8. Add sugar and salt.
9. When the tine is over, remove porridge from Instant Pot.
10. The dish is ready. Bon appetite!

Nutrition:

- Calories: 92
- Fat: 9g
- Carbohydrates: 12g
- Protein: 6g

Wheat porridge with fruits

Sweet porridge is cooked with whole grains of wheat, which denote eternal life and abundance. It is quite good source of inspiration, isn't it?

Prep time: 5 minutes
Cooking time: 55 minutes
Servings: 2

Ingredients:

- ½ cup rice
- ½ cup millet
- 4 ounce pumpkin
- 1 apple
- ¾ cup milk
- 2/3 cup water
- 1 teaspoon sugar
- 1 teaspoon salt

Directions:

1. Mix rice and millet.
2. Peel and dice apples.
3. Add apples to a mixture of rice and millet.
4. Grind pumpkin and add to rice and millet.
5. Place it into Instant Pot. Choose Porridge program.
6. Set a time at 35 minutes.
7. Add milk and water.
8. Add sugar and salt.
9. When the tine is over, remove porridge from Instant Pot.
10. The dish is ready. Bon appetite!

Nutrition:

- Calories: 92
- Fat: 9g
- Carbohydrates: 12g
- Protein: 6g

Pork in sweet and sour sauce

Tender meat with sweet and sour sauce will conquer your hearts. You've never tried such tenderness.

Prep time: 5 minutes
Cooking time: 55 minutes
Servings: 3

Ingredients:

- 1.5 pound pork
- 2 cloves of garlic
- 3 tablespoon flour
- 1 teaspoon salt
- 2 pickles
- 3 tablespoon vinegar
- 2 tablespoon sugar
- 4 ounce parsley
- 7 teaspoon farina
- 2 eggs
- 4 tablespoon water
- 1 teaspoon pepper

Directions:

1. Cut the pork into pieces.
2. Mix flour, sugar, and eggs. Knead a dough.
3. Add pepper and salt.
4. Dip pork into this mass.
5. Place pork into Instant Pot. Press the Power Button. Choose Meat/Stew program.
6. Set a time which is 55 minutes.
7. Take pieces of pork from Instant Pot. Let it cool down.
8. Make a sauce. Peel garlic and mix it with parsley, salt, pepper, and vinegar. Stir it well.
9. Garnish pork with this sauce. Bon appetite!

Nutrition:

- Calories: 265
- Fat: 18g
- Carbohydrates: 27g
- Protein: 13g

Escalope

Do you like veal? Then this dish is devoted for you. Escalope will satisfy the most demanding gourmet.

Prep time: 5 minutes
Cooking time: 45 minutes
Servings: 3

Ingredients:

- 1.5 pound veal
- 1 egg
- 1 ounce breadcrumbs
- ½ lemon
- 1 teaspoon salt
- 1 teaspoon pepper

Directions:

1. Tenderize a veal.
2. Season it with pepper and salt.
3. Beat an egg.
4. Sprinkle veal with egg.
5. Dredge meat with breadcrumbs.
6. Place veal into Instant Pot and press the Power Button. Choose Meat/Stew program.
7. Set a time which is 45 minutes.
8. Remove veal when the time is over. Let it cool down.
9. Escalopes are ready. You can sprinkle them with lemon.

Nutrition

- Calories: 235
- Fat: 16g
- Carbohydrates: 26g
- Protein: 11g

Pork with bananas

Do you like veal? Then this dish is devoted for you. Escalope will satisfy the most demanding gourmet.

Prep time: 5 minutes
Cooking time: 40 minutes
Servings: 4

Ingredients:

- 2 pound pork
- 4 bananas
- ½ lemon
- 1 teaspoon salt
- 1 teaspoon pepper

Directions:

1. Chop pork and tenderize each piece.
2. Season it with pepper and salt.
3. Sprinkle it with lemon
4. Peel bananas and cut it into small pieces.
5. Put 1 piece of bananas into center of each piece of pork.
6. Place pork into Instant Pot. Press the Power Button. Choose Meat/Stew program.
7. Arrange the needed time which is 40 minutes.
8. When the time is over, remove pork from Instant Pot.
9. Pork with bananas is ready. You can garnish it with potatoes or vegetables.

Nutrition:

- Calories: 270
- Fat: 18g
- Carbohydrates: 29g
- Protein: 13g

Pork in tomato sauce

One of the most delicious food combinations is pork and tomato sauce. It a decoration of any supper, isn't it?

Prep time: 10 minutes
Cooking time: 45 minutes
Servings: 3

Ingredients:

- 1 pound pork
- 8 ounce tomato paste
- 1/3 pound green peas
- 4 ounce farina
- 1 ounce soy sauce
- 1 white
- ¼ cup white wine
- 1 teaspoon sugar

Directions:

1. Slice pork.
2. Soft it in the mixture of white and farina.
3. Put it aside for 10 minutes.
4. Meanwhile mix tomato paste, soy sauce, whine wine and sugar. Stir it well.
5. Place pork into Instant Pot. Press the Power Button. Choose Meat/Stew program.
6. Set a time - 45 minutes.
7. When the time is over, remove pork from Instant Pot.
8. Sprinkle pork with a mixture of tomato paste, soy sauce, whine wine and sugar.
9. Pork in tomato sauce is ready. Garnish it with green peas.

Nutrition:

- Calories: 245
- Fat: 16g
- Carbohydrates: 25g
- Protein: 11g

Chinese pork

The feature of Chinese cuisine is its dishes from pork.
Check it by yourself!

Prep time: 5 minutes
Cooking time: 55 minutes
Servings: 1

Ingredients:

- 1/3 pound pork
- 1 white
- 1 ounce farina
- 4 teaspoon ginger
- 1 ounce soy sauce
- ¼ cup wine
- 1 teaspoon sugar
- 1 teaspoon salt
- 3 teaspoon water

Directions:

1. Slice pork
2. Mix farina, white and water. Stir it well.
3. Sprinkle fork with a mixture of farina, white and water.
4. Place pork into Instant Pot and press the Power Button. Choose Meat/Stew program.
5. Arrange the settings: set a time – 55 minutes.
6. When the time is over, remove the pork from Instant Pot. Let it cool down.
7. Meanwhile, mix ginger, soy sauce, wine, salt and sugar. Stir it well.
8. Sprinkle pork with a mixture of mixture of ginger, soy sauce, wine, salt and sugar.
9. The dish is ready. Bon appetite!

Nutrition:

- Calories: 253
- Fat: 18g
- Carbohydrates: 25g
- Protein: 12g

Pork with mushrooms

Tender pork with mushrooms is an ideal variant for supper, isn't' it?

Prep time: 5 minutes
Cooking time: 50 minutes
Servings: 2

Ingredients:

- 8 ounce pork
- 16 ounce mushrooms
- 1 egg
- 1 ounce farina
- 2 teaspoon soy sauce
- 2 teaspoon vodka
- 1 clove garlic
- 1 onion

Directions:

1. Mix eggs, farina and water. Stir it well.
2. Cut pork into small pieces.
3. Put pieces of fork into mixture of eggs, farina and water.
4. Cut mushrooms.
5. Place pork and mushrooms into Instant Pot. Press the Power Button. Choose Meat/Stew program.
6. Set a needed time which is 50 minutes.
7. When the time is over, remove pork from Instant Pot and let it cool down.
8. Meanwhile make a sauce. Peel and cut onion and garlic.
9. Mix soy sauce and vodka.
10. Add mixture of soy sauce and vodka to onion and garlic.
11. Sprinkle pork and mushrooms with this mass.
12. Pork with mushrooms is ready. Enjoy your meal!

Nutrition:

- Calories: 220
- Fat: 15g
- Carbohydrates: 23g
- Protein: 8g

Pork with potatoes and corn kernels

Pork + potatoes + corn kernels = tasty dish. You don't have to be a mathematician to understand this formula.

Prep time: 5 minutes
Cooking time: 55 minutes
Servings: 1

Ingredients:

- 1 cup corn kernels
- ¾ pound pork
- 1 onion
- 4 cups potatoes
- 1 cup water
- 1 cup milk
- 1 teaspoon salt
- 1 teaspoon pepper

Directions:

1. Chop pork.
2. Season it with pepper and salt.
3. Sprinkle pork with milk.
4. Peel potatoes and slice it.
5. Peel and cut onion.
6. Mix pork with potatoes, onion and corn kernels.
7. Place pork with vegetables into Instant Pot. Press the Power Button. Choose Meat/Stew program.
8. Arrange a time which is 55 minutes.
9. When the time is over, remove pork from Instant Pot. Let it cool down.
10. Pork with potatoes, onion and corn kernels is ready.

Nutrition:

- Calories: 245
- Fat: 17g
- Carbohydrates: 25g
- Protein: 10g

Austrian stuffed veal

Tender stuffed veal won't let you hungry. No so many ingredients, but so tasty dish is waiting for you!

Prep time: 5 minutes
Cooking time: 50 minutes
Servings: 2

Ingredients:

- 12 ounce veal
- 10 ounce pork
- 1 ounce Dutch cheese
- 3 eggs
- 1 ounce flour
- 1 ounce breadcrumbs
- 1 ounce butter
- 1 teaspoon salt
- 1 teaspoon pepper

Directions:

1. Chop veal.
2. Tenderize each piece.
3. Season each piece with pepper and salt.
4. Chop pork and put it into center of each veal pieces.
5. Grate cheese. Sprinkle meat with cheese.
6. Beat eggs.
7. Put flour and breadcrumbs into separate bowls.
8. Dip meat into egg and dredge with flour. Then dredge meat with breadcrumbs.
9. Place pieces of meat into Instant Pot. Press the Power Button. Choose Meat/Stew program.
10. Arrange the settings: time – 50 minutes.
11. When the time is over, remove meat from Instant Pot. Let it cool down.
12. You can garnish stuffed veal with any vegetables.

Nutrition:

- Calories: 230
- Fat: 15g
- Carbohydrates: 25g
- Protein: 10g

Chinese pieces

Chinese pieces are equivalent to meat balls. But it isn't so simple as it may seem. Try to cook it and check by yourself.

Prep time: 5 minutes
Cooking time: 50 minutes
Servings: 2

Ingredients:

- 1 pound mince
- 1 egg
- 1 chopped onion
- 1 tablespoon soy sauce
- 1 teaspoon sugar
- 1 teaspoon salt
- 1 tablespoon flour
- 4 ounce flour
- 1 yolk
- 2/3 cup milk,
- 4 ounce breadcrumbs.

Directions:

1. Mix mince, egg and onion. Stir it well.
2. Add salt, soy sauce and sugar.
3. Shape from mince small balls.
4. Beat yolk with milk.
5. Dip meatballs into egg.
6. Dredge them with flour and then dredge them with breadcrumbs.
7. Place meatballs into Instant Pot. Press the Power Button. Choose Meat/Stew program.
8. Arrange the settings: time – 50 minutes.
9. When the time is over, remove meatballs from Instant Pot. Let them cool down.
10. You can garnish meatballs with vegetables.

Nutrition:

- Calories: 180
- Fat: 13g
- Carbohydrates: 18g
- Protein: 11g

Chinese pieces

Chinese pieces are equivalent to meat balls. But it isn't so simple as it may seem. Try to cook it and check by yourself.

Prep time: 5 minutes
Cooking time: 55 minutes
Servings: 2

Ingredients:

- 1 pound mince
- 1 egg
- 1 chopped onion
- 1 tablespoon soy sauce
- 1 teaspoon sugar
- 1 teaspoon salt
- 1 tablespoon flour
- 4 ounce flour
- 1 yolk
- 2/3 cup milk
- 4 ounce breadcrumbs.

Directions:

1. Mix mince, egg and onion. Stir it well.
2. Add salt, soy sauce and sugar.
3. Shape small balls from mince.
4. Beat yolk with milk.
5. Dip meatballs into egg.
6. Dredge them with flour and then dredge with breadcrumbs.
7. Place meatballs into Instant Pot. Press the Power Button. Choose Meat/Stew program.
8. Arrange the settings: time – 55 minutes.
9. When the time is over, remove meatballs from Instant Pot. Let them cool down.
10. You can garnish meatballs with vegetables.

Nutrition:

- Calories: 180
- Fat: 13g
- Carbohydrates: 18g
- Protein: 11g

Cutlets "Ala-too"

These cutlets came from Kirghiz cuisine and hit us with harmony of tastes. Spicy but tender, these cutlets will drive anyone crazy.

Prep time: 5 minutes
Cooking time: 60 minutes
Servings: 1

Ingredients:

- 1/3 pound mutton
- 1 ounce milk
- 2 eggs
- 1 ounce butter
- 1 ounce breadcrumbs
- 1 ounce parsley
- 1 ounce flour
- ¼ cup milk
- 1 ounce olives
- 2 ounce green peas
- 2 ounce custard squashes
- 1 teaspoon pepper
- 1 teaspoon salt

Directions:

1. Add 1 yolk and milk to mince. Stir it well.
2. Make flat cakes from mince.
3. Put parsley into center of each flat cake. Seal with fork.
4. Beat an egg.
5. Dip each flat cake into egg.
6. Dredge with breadcrumbs.
7. Place flat cakes into Instant Pot. Pres the Power Button. Choose Meat/Stew program.
8. Set a time which is 60 minutes.
9. Remove flat cakes from Instant Pot when the time is over. Let them cool down.
10. Mix olives, custard squashes and green peas. Add salt and pepper.
11. Cutlets "Ala-too" are ready. Garnish them with olives, custard squashes and green peas.

Nutrition:

- Calories: 175
- Fat: 12g
- Carbohydrates: 20g
- Protein: 8g

Cutlets "Altay"

The dish looks very impressive, because it looks like a large cedar cone.

Prep time: 5 minutes
Cooking time: 50 minutes
Servings: 2

Ingredients:

- 1 pound pork
- 1 onion
- 1 ounce butter
- 1/3 pound breadcrumbs
- 1 teaspoon salt
- 1 teaspoon pepper

Directions:

1. Cut pork into pieces. Tenderize each piece.
2. Season with salt and pepper.
3. Chop onion.
4. Peel and beat eggs.
5. Mix eggs and onion.
6. Put mixture of eggs and onion into center of each piece of pork. Seal edges with a fork.
7. Place meat into Instant Pot. Pres the Power Button
8. Arrange the needed time which is 50 minutes.
9. When the time is over, remove meat from Instant Pot. Let it cool down.
10. Cutlets "Altay" are ready. You can garnish them with vegetables.

Nutrition:

- Calories: 220
- Fat: 14g
- Carbohydrates: 21g
- Protein: 10g

Tuzum-dulma

This meat dish will decorate any table and become a truly discovering. Tender beef with vegetables: what could be better?

Prep time: 5 minutes
Cooking time: 55 minutes
Servings: 1

Ingredients:

- 1 pound beef
- 1 onion
- 2 eggs
- 12 ounce potatoes
- 2 ounce tomatoes
- 4 ounce ketchup
- 1 teaspoon salt.

Directions:

1. Cut beef and tenderize each piece.
2. Season with salt.
3. Peel and cut onion.
4. Boil eggs.
5. Put boiled eggs into the center of each piece of beef.
6. Seal edges with a fork.
7. Place meat into Instant Pot. Pres the Power Button
8. Arrange settings: time – 55 minutes.
9. Remove beef from Instant Pot and let it cool down.
10. Meanwhile, peel and cut potatoes and tomatoes.
11. Mix potatoes and tomatoes. Add onion.
12. Tuzun-dulma is ready. Sprinkle it with ketchup and garnish with potatoes and tomatoes.

Nutrition:

- Calories: 210
- Fat: 13g
- Carbohydrates: 22g
- Protein: 9g

Kutaby

This dish is a combination of lamb and grenade which couldn't let you indifferent. Cook it and you won't let yourself stay calm.

Prep time: 5 minutes
Cooking time: 55 minutes
Servings: 2

Ingredients:

- 1 pound lamb
- 2 onions
- 2 cups wheat flour
- 4 ounce pomegranate seeds
- 2 tablespoons chopped greens
- 1/2 teaspoon cinnamon
- 1 teaspoon pepper
- 1 teaspoon salt

.**Directions:**

1. Sift a flour.
2. Beat eggs.
3. Mix flour, eggs and salt. Knead an elastic dough.
4. Roll out a dough and cut small circles.
5. Grind lamb and make mince.
6. Add onion and pomegranate seeds to mince.
7. Put mince into center of circles from dough. Seal edges with a fork.
8. Place meat into Instant Pot. Pres the Power Button. Choose Meat/Stew program.
9. Set a needed time which is 55 minutes.
10. When the time is over, remove lamb from Instant Pot. Let it cool down.
11. Kutaby is ready. Sprinkle it with cinnamon and serve with vegetables.

Nutrition:

- Calories: 180
- Fat: 11g
- Carbohydrates: 19g
- Protein: 7g

Lamb chops

This dish is a combination of lamb and grenade which could let you indifferent. Cook it and you won't let yourself to stay calm.

Prep time: 5 minutes
Cooking time: 60 minutes
Servings: 4

Ingredients:

- 2 pound mutton
- 10 onions
- 2 cups flour
- 5 cup milk
- 4 ounce butter
- 5 eggs
- 1 teaspoon salt
- 1 teaspoon pepper

Directions:

1. Cut mutton into pieces and tenderize each piece.
2. Season with salt and pepper.
3. Beat eggs.
4. Put 1 cup flour into bowl.
5. Dip mutton piece into egg. Then dredge it with flour.
6. Place pieces of mutton into Instant Pot. Press the Power Button. Choose Meat/Stew program.
7. Arrange the needed time – 60 minutes.
8. Meanwhile make pureed onions. Mix milk and 1 cup flour and stir it well.
9. Peel and slice onions.
10. Mix milk and 1 cup flour with onions. Add butter. Boil it for 20 minutes.
11. When the time is over, remove mutton from Instant Pot. Let it cool down.
12. Lamb chops are ready. Serve them with pureed onions. Bon appetite!

Nutrition:

- Calories: 230
- Fat: 13g
- Carbohydrates: 22g
- Protein: 9g

Meat Cheburek

This dish is a key of oriental cuisine, with a delicate piquant taste. Mince and dough – very simple but at the same time very delicious.

Prep time: 1 hour
Cooking time: 50 minutes
Servings: 2

Ingredients:

- 4 cup flour
- 1 cup water
- 1 teaspoon salt
- 1 pound mince
- 1/3 pound onions

Directions:

1. Mix flour and water. Stir it well. As a result, it should be a steep dough
2. Put a dough aside for 1 hour.
3. Chop onions.
4. Add onions to mince.
5. Roll out a dough and shape small flat cakes.
6. Put mince in the center of each flat cake. Seal the edges with a fork.
7. Place flat cakes into Instant Pot. Press the Power Button. Choose Meat/Stew program.
8. Arrange settings: time – 50 minutes.
9. Remove flat cakes from Instant Pot and let them cool down.
10. Meet chebureks are ready. Enjoy your meal!

Nutrition:

- Calories: 260
- Fat: 15g
- Carbohydrates: 24g
- Protein: 10g

Stuffed breast of veal

Everything is perfect in these word combinations – every word represents the harmony of tastes. Enjoy a tender stuffed veal!

Prep time: 1 hour
Cooking time: 55 minutes
Servings: 4

Ingredients:

- 2 pound brisket meat
- 2 eggs
- 4 ounce breadcrumbs
- 2 ounce butter
- 3 ounce cream
- 4 ounce almonds
- 1 lemon
- ¼ cup cognac
- 1 teaspoon salt
- 1 teaspoon pepper

Directions:

1. Wash the brisket meat.
2. Cut brisket meat into pieces. The lower, wider part is carefully cut from the bone in such a way that a pocket is obtained.
3. Season with salt.
4. Mix eggs, bread, and almonds.
5. Then add the juice of lemon, salt and pepper. Stir it well.
6. The meat pocket is filled with this mass; after which it is sewn.
7. Place meat pocket into Instant Pot. Press the Power Button. Choose Meat/Stew program.
8. Arrange the settings: time – 55 minutes.
9. Remove meat pocket from Instant Pot. Let them cool down.
10. Make a sauce: mix cognac, cream and spices. Stir it well.
11. Stuffed breast of veal is ready. Sprinkle it with a sauce. Have a nice meal!

Nutrition:

- Calories: 245
- Fat: 13g
- Carbohydrates: 23g
- Protein: 10g

Cutlets "Holiday"

Cutlets that imply holidays cannot be omitted. Arrange holidays for yourself!

Prep time: 5 minutes
Cooking time: 1 hour
Servings: 2

Ingredients:

- 1 pound beef
- 5 cloves of garlic
- 1/3 pound butter
- 3 eggs
- 4 ounce flour
- ½ cup milk
- 4 ounce breadcrumbs
- 2 ounce parsley
- 1 teaspoon salt
- 1 teaspoon pepper

Directions:

1. Cut meat into pieces.
2. Tenderize it.
3. Season with salt and pepper.
4. Rub with garlic.
5. In the middle of each piece put a piece of butter.
6. Give the cutlets an elongated shape.
7. Mix eggs, milk and flour. Beat it well.
8. Dip meat into a mixture of eggs, milk and flour.
9. Dredge with breadcrumbs.
10. Place meat into Instant Pot. Press the Power Button. Choose Meat/Stew program.
11. Set a time at 1 hour.
12. When the time is over, remove meat from Instant Pot.
13. Cutlets "Holiday" are ready. Sprinkle with parsley before serving.

Nutrition:

- Calories: 225
- Fat: 12g
- Carbohydrates: 23g
- Protein: 10g

Kovurma of Chuchvar

Strange name conceals the delicious dish which consists of spicy meat and gentle dough. Cook it and check by yourself!

Prep time: 10 minutes
Cooking time: 55 minutes
Servings: 2

Ingredients:

- 4 cup flour
- 1 egg
- ½ cup water,
- 2 teaspoon salt.
- 1 pound mutton
- 4 onions
- 1 teaspoon pepper

Directions:

1. Dissolve salt in 1 tablespoon warm water.
2. Beat the egg.
3. Add flour and remaining water.
4. Knead a steep dough. Put it aside for 10 minutes.
5. Roll the dough into a thin layer and cut it into squares of 5x5 cm.
6. Put the hem into squares of the dough. Sprinkle with flour and cover with a napkin.
7. Make a mince from mutton.
8. Add chopped onion, salt and pepper.
9. Put mince into center of squares of dough. Seal the edges with a fork.
10. Place meat into Instant Pot. Press the Power Button. Choose Meat/Stew program.
11. Arrange the settings: time – 55 minutes.
12. Remove meat, when the time is over. Let it cool down.
13. Kovurma of Chuchvar is ready. Bon appetite!

Nutrition:

- Calories: 245
- Fat: 14g
- Carbohydrates: 27g
- Protein: 9g

Spanish tenderized steak

Temperamental Spain on your table! Try these delicious pork chops and deep into Spanish atmosphere!

Prep time: 5 minutes
Cooking time: 45 minutes
Servings: 2

Ingredients:

- 1 pound pork
- 4 ounce breadcrumbs
- 2 cloves of garlic
- 4 sweet peppers
- 4 ounce prunes
- greenery
- 1 teaspoon salt
- 1 teaspoon pepper

Directions:

1. Cut pork.
2. Tenderize each piece.
3. Season with salt and pepper.
4. Mix breadcrumbs, garlic, greenery, salt, pepper. Make a homogeneous mass.
5. Rub meat with this mass.
6. Place meat into Instant Pot. Press the Power Button. Choose Meat/Stew program.
7. Arrange the settings: time – 45 minutes.
8. Meanwhile, prunes are scalded with boiling water. Remove kernels.
9. When the time is over, remove meat from Instant Pot.
10. Spanish tenderized steak is ready. Serve it on a hot plate with pepper and prunes.

Nutrition:

- Calories: 215
- Fat: 12g
- Carbohydrates: 17g
- Protein: 9g

Chevapchichi

The simplest in the world! As easy as never. Mean and spices – nothing extra.

Prep time: 5 minutes
Cooking time: 55 minutes
Servings: 2

Ingredients:

- 1 pound beef
- 1 onion
- 1 teaspoon pepper
- 1 teaspoon salt
- 1 teaspoon vegetable oil

Directions:

1. Make a mince from a beef.
2. Season it with pepper and salt.
3. Knead it well.
4. Shape flat sausages.
5. Smear them with a small amount of vegetable oil
6. Place sausages into Instant Pot and press the Power Button. Choose Meat/Stew program.
7. Arrange the needed time which is 55 minutes.
8. When the time is over, remove meat from Instant Pot. Let it cool down.
9. Chevapchichi is ready. Serve it with vegetables.

Nutrition:

- Calories: 215
- Fat: 12g
- Carbohydrates: 17g
- Protein: 9g

Cutlets "Tiraspol"

Of course, Tiraspol is a beautiful city, nevertheless, Tiraspol is also a delicious meat dish. The most tender in the world. Do you believe in that?

Prep time: 5 minutes
Cooking time: 50 minutes
Servings: 1

Ingredients:

- 8 ounce pork
- 1/3 pound pork liver
- 1 onion
- 2 egg
- 4 ounce butter
- 8 ounce melted lard
- spices
- 1 teaspoon salt,
- 1 teaspoon pepper

Directions:

1. Cut pork into slices across the fibers.
2. Tenderize each piece.
3. Season with salt and spices.
4. Fry pork liver and boil the eggs.
5. Cut liver and eggs into strips. Add onion, butter and pepper. Stir it well – mince should be viscid.
6. Put mince in the middle of each slice of pork. Seal the edges.
7. Place meat into Instant Pot. Press the Power Button. Choose Meat/Stew program.
8. Set a time at 50 minutes.
9. When the time is over, remove meat from Instant Pot. Let it cool down.
10. Cutlets “Tiraspol” are ready. Bon appetite!

Nutrition:

- Calories: 215
- Fat: 12g
- Carbohydrates: 17g
- Protein: 9g

Moldavian meat

Meat that is full of spices couldn't disappoint you. Moldavian cuisine is appropriate for meat satisfaction.

Prep time: 5 minutes
Cooking time: 15 minutes
Servings: 1

Ingredients:

- 8 ounce beef tenderloin
- 4 cloves of garlic
- 4 cup vegetable oil
- 8 ounce parsley
- 8 ounce cloves
- 2 teaspoon salt
- 2 teaspoon black and red ground pepper

Directions:

1. Cut meat into slices. The thickness is approximately 0.02 inches.
2. Tenderize each piece of meat.
3. Season it with black and red pepper, crushed cloves, chopped garlic, salt.
4. Roll pastry. Seal the edges.
5. Place meat into Instant Pot. Press the Power Button. Choose Meat/Stew program.
6. Set a time at 15 minutes.
7. When the time is over, remove meat from Instant Pot. Let it cool down.
8. Moldavian meat is ready. Sprinkle it with parsley. Enjoy your meals!

Nutrition:

- Calories: 195
- Fat: 14g
- Carbohydrates: 21g
- Protein: 10g

Collar of brawn

Find a true source of power! Tasty pork is a perfect variant, isn't it?

Prep time: 12 hours
Cooking time: 55 minutes
Servings: 4

Ingredients:

- 3 pound pork fillet
- 4 cloves of garlic
- 1 pound fat
- 1 branch of thyme
- 1 teaspoon salt
- 1 teaspoon pepper

Directions:

1. Season meat with salt, pepper and rub with garlic.
2. Make a roll and fasten it with a thread. Put it aside for 12 hours.
3. After that, place meat into Instant Pot. Press the Power Button. Choose Meat/Stew program.
4. Arrange the settings: time – 55 minutes.
5. When the time is over, remove roll from Instant Pot. Let it cool down.
6. Collar of brawn is ready. Sprinkle it with a branch of thyme Serve it with any vegetables.

Nutrition:

- Calories: 220
- Fat: 13g
- Carbohydrates: 24g
- Protein: 10g

Veal escallops in Corsican style

Very and very delicious. Pamper yourself with tender veal and get acquainted with Corsican style.

Prep time: 5 minutes
Cooking time: 50 minutes
Servings: 4

Ingredients:

- 4 veal escallops
- 4 slices ham
- 1 onion
- ¾ cup Madeira
- 1 teaspoon basil
- 1 teaspoon thyme
- 1 teaspoon sage
- 1 ounce flour
- 2 tablespoon olive oil
- 4 ounce butter
- 2 teaspoon salt
- 2 teaspoon pepper

Directions:

1. Tenderize the escallops.
2. Season it with chopped herbs.
3. Put slices of ham on top of the escallops. Wrap them so that the ham must be inside.
4. Dredge meat with flour.
5. Place meat into Instant Pot. Press the Power Button. Choose Meat/Stew program. Choose Meat/Stew program. Set a time at 50 minutes.
6. Remove meat when the time is over. Let it cool down.
7. Mix basil, thyme, olive oil and Madeira. Stir it well.
8. Sprinkle meat with a mixture of basil, thyme, olive oil and Madeira.
9. Veal escallops in *Corsican style is ready. Bon appetite!*

Nutrition:

- Calories: 210
- Fat: 10g
- Carbohydrates: 18g
- Protein: 7g

Rabbit with parsley

Hurry up to catch a tasty rabbit! And when you catch him, it will be hard to let him go…

Prep time: 5 minutes
Cooking time: 55 minutes
Servings: 4

Ingredients:

- Rabbit carcass
- 4 slices of smoked bacon
- 2 ounce ham
- 2 pound parsley
- 1 tablespoon butter
- ¾ cup vegetable oil
- 1 teaspoon salt
- 1 teaspoon pepper.

Directions:

1. Cut each leg of the rabbit in half.
2. Cut the back of the carcass into 4 parts.
3. Wash parsley.
4. Tenderize each part of rabbit.
5. Season it with salt and pepper.
6. Put slices of bacon and ham into the center of pieces of rabbit. Seal the edges.
7. Sprinkle meat with vegetable oil and butter.
8. Place meat into Instant Pot. Press the Power Button. Choose Meat/Stew program.
9. Arrange the settings: time – 55 minutes.
10. When the time is over, remove roll from Instant Pot. Let it cool down.
11. Rabbit is ready. Serve it with parsley. Have a nice meal!

Nutrition:

- Calories: 170
- Fat: 15g
- Carbohydrates: 21g
- Protein: 9g

Cold boiled pork “in 2 seconds”

Very quickly and very tasty! This cold boiled pork is ready to impress you. Are you ready?

Prep time: 20 minutes
Cooking time: 50 minutes
Servings: 4

Ingredients:

- 2 pound pork
- 2 tablespoon tomato paste
- 4 ounce cream
- 1 cup red dry wine
- 1 tablespoon cumin.
- 1 tablespoon coriander
- 1 tablespoon paprika sweet
- 3 tablespoon olive oil

Directions:

1. Make holes in the pork with the help of knife.
2. Cut carrot and put pieces into these holes.
3. Also, cut garlic and add garlic to carrot.
4. Smear meat vegetable oil (salt, mustard and spices. Put it aside for 20 minutes.
5. Place pork into Instant Pot. Press the Power Button. Choose Meat/Stew program.
6. Arrange the needed settings: time – 50 minutes.
7. When the time is over. Remove pork from Instant Pot. Let it cool down.
8. Cold boiled pork is ready. Serve with any vegetables.

Nutrition:

- Calories: 170
- Fat: 15g
- Carbohydrates: 21g
- Protein: 9g

Pork Roxas

It is one of the most popular Spanish dishes. It is a tender meat with delicious sauce which deserves the right to be one of the best ones. Do you agree?

Prep time: 5 minutes
Cooking time: 55 minutes
Servings: 4

Ingredients:

- 2 pound pork
- 1 carrot
- 6 cloves of garlic
- 2 tablespoon vegetable oil
- 2 tablespoon mustard
- 1 tablespoon spices.
- 1 tablespoon salt

Directions:

1. Cut pork into thin long blocks.
2. Mix the tomato paste with cumin, coriander and paprika. Add the cream. Stir it well.
3. Place pork into Instant Pot. Press the Power Button. Choose Meat/Stew program.
4. Arrange the needed settings: temperature – 180 degrees, time – 20 minutes.
5. Then add mixture of tomato paste with cumin, coriander and paprika to pork.
6. Change the program to Slow Cook.
7. Set a time at 55 minutes.
8. When the time is over, remove pork from Instant Pot. Let it cool down.
9. Cold boiled pork is ready. Serve with any vegetables.

Nutrition:

- Calories: 220
- Fat: 17g
- Carbohydrates: 21g
- Protein: 12g

Cutlets with potatoes

Sounds not so exquisitely, nevertheless, the taste is amazing! Cutlets and potatoes are ideal combination of tastes.

Prep time: 5 minutes
Cooking time: 40 minutes
Servings: 2

Ingredients:

- 1 pound mince
- 7 potatoes
- 1 onion
- 1 clove of garlic
- 1 tablespoon vegetable oil
- 1 tablespoon salt

Directions:

1. Add finely chopped onions and grated garlic to mince.
2. Season with salt.
3. Shape cutlets from mince.
4. Place cutlets into Instant Pot. At the bottom, we pour a lot of oil. Press the Power Button.
5. Choose program Meat/Stew.
6. Set a needed time which is 40 minutes.
7. Meanwhile, peel and slice potatoes.
8. After 20 minutes, add potatoes to meat in Instant Pot.
9. When the time is over remove meat and potatoes from Instant Pot. Let it cool down.
10. The dish is ready. Have a nice meal!

Nutrition:

- Calories: 240
- Fat: 16g
- Carbohydrates: 23g
- Protein: 10g

Domlama

True treasure of Uzbek cuisine! It is very easy to cook this useful and nutritious dish. Let's go!

Prep time: 5 minutes
Cooking time: 50 minutes
Servings: 2

Ingredients:

- 1 pound beef
- 8 ounce fat
- 2 onions
- 1 carrot
- 2 tomatoes
- 1 Bulgarian pepper
- 1 eggplant
- 1 tablespoon salt
- spices (pepper, ground coriander, zira)
- 10 cloves of garlic
- parsley
- White cabbage (several leaves)

Directions:

1. Place fat at the bottom of Instant Pot.
2. Peel and slice onion.
3. Put onions above the fat.
4. Cut beef into pieces.
5. Season with salt and spices.
6. Put beef above onions.
7. Peel and cut carrot.
8. Put carrot above the beef.
9. Peel and cut tomatoes.
10. Put tomatoes above the carrot
11. Peel and cut Bulgarian pepper
12. Put Bulgarian pepper above the tomatoes
13. Peel and cut Bulgarian eggplant and put it above Bulgarian pepper
14. Then sprinkle with chopped herbs and cover with cabbage leaves.
15. Press the Power Button. Choose Meat/Stew Program.
16. Arrange the needed time which is 50 minutes.
17. When the time is over, remove it from Instant Pot.
18. Domlama is ready. Bon appetite!

Nutrition:

- Calories: 260
- Fat: 16g
- Carbohydrates: 25g
- Protein: 12g

Haricot

It isn't as easy as it may seem. The key feature of this dish – plum jam. Interested?

Prep time: 5 minutes
Cooking time: 1 hour
Servings: 4

Ingredients:

- 2 pound pork
- 2 tablespoon plum jam
- ½ cup wine red dry
- 1 onion
- 3 cloves of garlic
- 1 carrot
- 8 ounce celiac celery
- 4 ounce dry mushrooms
- 5 tablespoon olive oil
- 2 cup water
- 1 tablespoon flour wheaten
- 1 tablespoon salt
- 1 tablespoon black ground pepper
- 1 tablespoon cumin.
- 1 tablespoon paprika sweet

Directions:

1. Dice pork. Season it with salt, black pepper, and sweet paprika. Stir it well.
2. Dice vegetables: carrots, celery, garlic and onion.
3. Place vegetables into Instant Pot. Add olive oil.
4. Add pork to vegetables.
5. Choose program Meat/Stew.
6. Set the needed time at 45 minutes.
7. When the time is over, add mushrooms. Add additional time – 15 minutes.
8. Remove meat from Instant Pot. Sprinkle with flour and wine.
9. Haricot is ready. Serve with plum jam.

Nutrition:

- Calories: 210
- Fat: 14g
- Carbohydrates: 23g
- Protein: 9g

<u>Cutlets "Sophie" with broccoli</u>

It is quite easy recipe that will surprise you by its harmony of tastes. Delicious cutlets with a minimum amount of fat.

Prep time: 5 minutes
Cooking time: 50 minutes
Servings: 2

Ingredients:

- 1 pound veal
- 1/3 ham
- 2 boiled eggs
- 4 tablespoon cream
- 3 tablespoon butter L.
- 1 pound broccoli
- ½ lemon
- 1 tablespoon salt
- 1 tablespoon pepper

Directions:

1. Cut veal and tenderize each piece.
2. Season with salt and pepper. Sprinkle with lemon juice.
3. Put slices of ham and half of egg onto each pieces of veal.
4. Seal the edges of veal pieces.
5. Place meat inti Instant Pot. Press the Power Button.
6. Choose program Meat/Stew and set a time at 50 minutes.
7. After 20 minutes, add broccoli to Instant Pot.
8. When the time is over remove meat and broccoli from Instant Pot,
9. Cutlers "Sophie" are ready. Enjoy your meal!

Nutrition:

- Calories: 230
- Fat: 15g
- Carbohydrates: 24g
- Protein: 11g

Beef Stroganoff

Juicy beef that is easy to cook is a recipe for happiness. This dish will empower you with lots of positive emotions.

Prep time: 20 minutes
Cooking time: 40 minutes
Servings: 2

Ingredients:

- 1 pound beef
- 1 onion.
- 4 cloves of garlic
- 1 teaspoon salt
- 1 teaspoon black pepper
- 2 tablespoon tomato paste.
- 3 tablespoon sour cream
- 2 tablespoon mustard

Directions:

1. Cut the meat across the fibers into small pieces
2. Tenderize each piece.
3. Then, smear each piece with mustard. Put aside for 20 minutes.
4. Place meat into Instant Pot. Press the Power Button.
5. Choose program Meat/Stew and set a time at 40 minutes.
6. Chop garlic and onion and add to meat.
7. After 20 minutes, mix sour cream and tomato paste. Add to meat.
8. Add water.
9. When the time is over, remove it from Instant Pot.
10. Beef Stroganoff is ready. You can garnish it with rice.

Nutrition:

- Calories: 245
- Fat: 16g
- Carbohydrates: 27g
- Protein: 12g

Spareribs

1,2,3 and dish is ready! Very tasty and very nutritious.

Prep time: 2 hours
Cooking time: 40 minutes
Servings: 4

Ingredients:

- 3 pound pork
- 2 tomatoes
- 1 onion
- 1 teaspoon black pepper
- 4 tablespoon soy sauce
- 3 tablespoon tomato paste
- 3 tablespoon olive oil

Directions:

1. Wash and cut ribs along the bones.
2. Place ribs in a large bowl of.
3. Dice tomatoes and add to ribs.
4. Cut onion and add to ribs.
5. Mix soy sauce, tomato paste and olive oil. Stir it well.
6. Sprinkle meat with a mixture of soy sauce, tomato paste and olive oil. Put it into fridge for 2 hours.
7. After that, place meat into Instant Pot. Press the Power Button.
8. Choose program Meat/Stew and set a time at 40 minutes.
9. When the time is over, remove it from Instant Pot.
10. Spareribs are ready. Bon appetite!

Nutrition:

- Calories: 215
- Fat: 13g
- Carbohydrates: 24g
- Protein: 9g

Beef in beer

Juicy, tender beef with a thick, fragrant gravy. Men will be very happy about a meat dish, and women will not refuse from pampering yourselves.

Prep time: 5 minutes
Cooking time: 40 minutes
Servings: 2

Ingredients:

- 1 pound beef
- 1 onion
- 1 tablespoon butter
- 1 tablespoon vegetable oil
- 2 ounce wheat flour
- 2 cup beer
- 1 tablespoon mustard
- 2 slices bread
- 1 tablespoon Salt
- 1 tablespoon Black pepper

Directions:

1. Dice meat.
2. Dice onion.
3. Mix dice and onions.
4. Season beef with pepper and salt.
5. Dredge meat pieces with flour.
6. Place them into Instant Pot. Press the Power Button.
7. Choose program Meat/Stew and set a time at 40 minutes.
8. While meat is cooking prepare a sauce.
9. Mix butter, vegetable oil, mustard and beer. Stir it well.
10. After 20 minutes, sprinkle meat with this sauce.
11. When the time is over, remove beef from Instant Pot. Let it cool down.
12. Serve beef in beer with bread. Bon appetite!

Nutrition:

- Calories: 234
- Fat: 12g
- Carbohydrates: 14g
- Protein: 6g

Beef goulash with chick peas and vegetables

A nutritious and healthy dish for autumn and winter cold. Tender, melting meat with a tomato gravy and delicious chickpeas will impress you!

Prep time: 12 hours
Cooking time: 55 minutes
Servings: 4

Ingredients:

- 1 cup chickpeas (raw)
- 2 pound beef
- 2 onions
- 1 Bulgarian pepper
- 1 pound tomatoes
- 1 tablespoon tomato paste
- 3 tablespoon vegetable oil
- 1 tablespoon paprika sweet
- 1 chili pepper
- Thyme or basil
- 1 tablespoon salt

Directions:

1. Soak chickpeas in plenty of water for at least 12 hours.
2. Cut beef into pieces.
3. Peel and cut onion.
4. Make a sauce: mash peeled tomatoes with the tomato paste, Bulgarian pepper, and chili pepper. Add spices and herbs.
5. Place beef into Instant Pot. Press the Power Button.
6. Place chickpeas above meat.
7. Choose program Meat/Stew and set a time at 55 minutes.
8. When the time is over, remove meat from Instant Pot.
9. Beef goulash with chick peas and vegetables is ready. Enjoy your meal!

Nutrition:

- Calories: 252
- Fat: 14g
- Carbohydrates: 24g
- Protein: 9g

Iranian mutton

There is something special in this dish. Try to cook it and discover the lamb which you've never imagined to be.

Prep time: 12 minutes
Cooking time: 40 minutes
Servings: 2

Ingredients:

- 1 pound mutton
- 2 apples
- 1 onion
- 3 cloves of garlic
- 1 tablespoon flour
- 1 tablespoon curry
- 1 tablespoon salt
- 1 tablespoon pepper
- ½ lemon

Directions:

1. Dice mutton.
2. Season mutton with salt and pepper.
3. Peel and chop onions and garlic
4. Place onions, garlic and curry into Instant Pot. Press the Power Button.
5. Choose program Slow Cook. Set a time at 15 minutes.
6. Then put mutton into Instant Pot. Choose program Meat/Stew and set a time at 40 minutes.
7. Peel and slice apples.
8. Put apples into Instant Pot after 20 minutes.
9. When the time is over remove it from Instant Pot.
10. Serve the dish hot. Sprinkle meat with lemon juice.
11. Iranian mutton is ready. Have a nice meal!

Nutrition:

- Calories: 230
- Fat: 14g
- Carbohydrates: 24g
- Protein: 9g

Pork "24 hours"

It isn't difficult to guess that this dish requires many time. Despite its durability pork is cooked almost without your participation. Surprised? We too.

Prep time: 12 hours
Cooking time: 5 hours
Servings: 4

Ingredients:

- 2.5 pound pork
- 1 orange
- 2 tablespoon honey
- 4 tablespoon soy sauce
- 3 cloves of garlic
- 1 tablespoon cumin
- 1 tablespoon coriander
- 1 tablespoon red hot pepper
- 1 tablespoon marjoram
- 1 tablespoon chaiber
- 1 tablespoon rosemary
- 1 tablespoon pepper
- 2 tablespoon vegetable oil
- 1 tablespoon salt

Directions:

1. Wash and cut pork.
2. Peel and grate orange.
3. Mix honey, oranges garlic, soy sauce and vegetable oil. Stir it well.
4. Sprinkle pork with a mixture of honey, oranges garlic, soy sauce and vegetable oil. Put it into the fridge for 12 hours.
5. Mix all spices. Stir it well.
6. Sprinkle pork with spices.
7. Place meat into Instant Pot. Choose program Meat/Stew and set a time at 1 hour.
8. When the time is over, choose another program – Keep Warm.
9. Set a time at 4 hours.
10. When the time is over, remove pork from Instant Pot.
11. Pork is ready. It would be better to garnish it with rice.

Nutrition:

- Calories: 215
- Fat: 12g
- Carbohydrates: 22g
- Protein: 9g

Meat Rolls

It is a recipe for juicy and tasty rolls stuffed with bacon, cheese and sun-dried tomatoes! Sounds inspiring, isn't' it?

Prep time: 5 minutes
Cooking time: 1 hour
Servings: 2

Ingredients:

- 1 pound pork
- 2 ounce bacon
- 1 ounce butter
- 2 tablespoon vegetable oil
- ½ cup broth
- ½ cup cream
- 2 tablespoon salt
- 1/3 pound dried tomatoes
- 1/3 pound parmesan
- 2 tablespoon pepper

Directions:

1. Cut pork into pieces.
2. Tenderize each peace.
3. Season with pepper.
4. Slice bacon.
5. Grate cheese.
6. Put bacon and cheese onto pieces of pork.
7. Put on the edges of pork dried tomatoes.
8. Make rolls.
9. Each roll dredge with flour.
12. Place rolls into Instant Pot. Choose program Meat/Stew and set a time at 1 hour.
10. Meanwhile make a sauce.
11. Mix cream, broth, vegetable oil and butter. Stir it well.
12. When the time is over, remove rolls from Instant Pot. Let them cool down.
13. Serve meat rolls with a sauce. Enjoy your meal!

Nutrition:

- Calories: 245
- Fat: 14g
- Carbohydrates: 23g
- Protein: 10g

Pork in apple-ginger marinade

This dish makes everyone's mouth water. You cannot refuse from it. Just little piece if this dish will drive you crazy.

Prep time: 4 hours
Cooking time: 1 hour
Servings: 2

Ingredients:

- 1 pound pork
- 3 onions
- 1 Bulgarian pepper
- 1 carrots
- 1 apple
- 1 ginger
- 6 tablespoon soy sauce
- 5 cloves of garlic
- 1 tablespoon chili pepper
- 1 tablespoon black pepper.
- 1 tablespoon brown sugar
- 2 tablespoon honey
- 2 tablespoon Sesame oil

Directions:

1. Slice pork into thin pieces.
2. Slice carrot.
3. Cut onions into 4 pieces.
4. Cut pepper into pieces.
5. Make marinade.
6. Peel and grate apples.
7. Grate ginger.
8. Cut chili pepper into small pieces.
9. Mix all these ingredients and stir it well.
10. Marinade pork and put it into the fridge for 4 hours.
11. Place pork, onions, carrot and pepper into Instant Pot. Choose program Meat/Stew and set a time at 1 hour.
12. When the time is over, remove meat from Instant Pot.
13. The dish is ready. You can serve it with rice or potatoes.

Nutrition:

- Calories: 235
- Fat: 13g
- Carbohydrates: 22g
- Protein: 11g

Burgundy Beef

It is a well-known fat that meat and wine is a perfect combination of tastes. Try this delicate and tender meat, with a pleasant piquant note.

Prep time: 4 hours
Cooking time: 1 hour
Servings: 2

Ingredients:

- 1 tablespoon black pepper
- 1 tablespoon salt
- ¼ cup cognac
- ¾ cup red semi-sweet wine
- 2 red onions
- 1/3 pound champignons
- 1 carrot
- 1 pound beef
- 2 cloves of garlic.

Directions:

1. Dice beef.
2. Slice onions.
3. Cut carrot.
4. Chop champignons
5. Place carrot into Instant Pot.
6. Place beef above the carrot.
7. Cut garlic and put it above the beef.
8. Place onions and champignons above the meat.
9. Season with salt and pepper.
10. Add wine and cognac.
11. Choose program Meat/Stew and set a time at 1 hour.
12. When the time is over, remove the dish from Instant Pot.
13. Burgundy beef is ready. Bon appetite!

Nutrition:

- Calories: 195
- Fat: 11g
- Carbohydrates: 17g
- Protein: 8g

Rabbit with vegetables and bulgar

It is an unusual combination of usual rabbit and unusual bulgar. Nevertheless, this dish possesses a bright and unforgettable taste. Try to check it by yourself.

Prep time: 5 minutes
Cooking time: 1 hour and 30 minutes
Servings: 4

Ingredients:

- 2.5 pound rabbit
- 1 carrot
- 3 Bulgarian peppers (red, green, yellow).
- 1 onion
- 5 tablespoon olive oil.
- 1 cup bulgar - 1 stack.
- 1 tablespoon salt
- 1 tablespoon pepper
- 1 tablespoon basil

Directions:

1. Cut rabbit into pieces the size of which is approximately 2x2 inches.
2. Season rabbit with salt and pepper.
3. Cut carrot and Bulgarian peppers.
4. Place rabbit into Instant Pot. Choose program Meat/Stew and set a time at 1 hour.
5. Place vegetables above the rabbit.
6. When the time is over, add bulgar and change the program to Low Cooking.
7. Set a time – 30 minutes.
8. When the time is over, remove dish from Instant Pot.
9. Rabbit is ready. You can garnish it with vegetables.

Nutrition:

- Calories: 205
- Fat: 12g
- Carbohydrates: 19g
- Protein: 10g

Spring meat soufflé

It is e very simple recipe of meat suffle with vegetables. By the way, this dish will help us to be slim and beautiful by the beginning of the summer season.

Prep time: 5 minutes
Cooking time: 50 minutes
Servings: 2

Ingredients:

- 1 ounce corn
- ½ cup milk
- 1 ounce green peas
- 1 tablespoon oatmeal flakes.
- 1 egg
- 1 pound veal

Directions:

1. Grind the veal.
2. Separate white from yolk.
3. Mix veal, yolk, milk and oatmeal flakes. Stir it well.
4. Whip the white and add to meat.
5. Stir it well.
6. Place meat into silicone molds and put them into Instant Pot. Choose program Meat/Stew
7. Set a time at 50 minutes.
8. When the time is over, remove meat from Instant Pot.
9. Souffle is ready. Serve it with peas. Bon appetite!

Nutrition:

- Calories: 175
- Fat: 11g
- Carbohydrates: 26g
- Protein: 7g

Meat stew with cheese balls

Stew requires a long cooking time, but as a result you'll get unusually tasty dish! Tender meat, mushrooms, sweetish vegetables, cheese balls - all of these are in a thick, rich sauce.

Prep time: 5 minutes
Cooking time: 1 hour
Servings: 2

Ingredients:

- 1 pound beef
- 1 onion
- 2 tablespoon vegetable oil
- 3 tablespoon flour
- 2 cup water
- 3 tablespoon soy sauce
- 1 tablespoon tomato paste
- 2 carrots
- 4 ounce mushrooms
- Laurel leaf
- 1 tablespoon mustard
- 1 tablespoon sugar
- 1 tablespoon pepper
- 4 ounce cheese
- 1 tablespoon salt
- Parsley
- 1 clove of garlic
- 2 tablespoon cream.
- 1 egg of chicken

Directions:

1. Dice meat. Season with salt and pepper.
2. Mix tomato paste, soy sauce, sugar, pepper, mustard and water - add to the meat. Stir it well.
3. Cut mushrooms and carrot and add to the meat.
4. Place meat into Instant Pot. Choose program Meat/Stew. Set a time at 1 hour.
5. Mix the egg, cream, grated cheese, chopped garlic and parsley, salt, and flour. Knead a dough.
6. Shape small balls from a dough.
7. After 40 minutes, place these balls into Instant Pot
8. When the time is over, remove the dish from Instant Pot.
9. Meat stew with cheese balls is ready. Serve it hot.

Nutrition:

- Calories: 175
- Fat: 11g
- Carbohydrates: 26g
- Protein: 7g

Pork ribs with potatoes

Very easy. Very tasty. Always eaten. Never left.

Prep time: 5 minutes
Cooking time: 50 minutes
Servings: 2

Ingredients:

- 1 pound pork ribs
- 2 pound potatoes
- 1 onion.
- 2 cloves of garlic
- dill and parsley
- spices
- 1 tablespoon salt
- 1 tablespoon white pepper
- 1 tablespoon vegetable oil

Directions:

1. Season pork with spices and salt.
2. Cut potatoes.
3. Cut onions.
4. Chop garlic.
5. Put everything into Instant Pot. Choose program Meat/Stew. Set a time at 1 hour.
6. When the time is over, remove the dish from Instant Pot.
7. Pork ribs with potatoes are ready. Bon appetite!

Nutrition:

- Calories: 225
- Fat: 16g
- Carbohydrates: 26g
- Protein: 9g

Hun-shao-jo

Of course, it is a dish of Asian cuisine. This dish implies brisket pork that is cooked with the skin. The skin becomes very soft, delicate, and the meat is delicious and juicy.

Prep time: 5 minutes
Cooking time: 1 hour 15 minutes
Servings: 2

Ingredients:

- 1 pound brisket pork
- 1 onion
- 1 ginger.
- 2 cloves of garlic
- Sternanis
- 1 tablespoon cinnamon
- 1 ounce sugar
- 2 tablespoon soy sauce
- 5 tablespoon vegetable oil.
- 3 tablespoon red semi-dry wine
- 1 tablespoon salt
- 2 tablespoon paprika sweet.
- 4 cups water

Directions:

1. Dice brisket pork
2. Place meat into Instant Pot. Choose program Meat/Stew. Set a time at 15 minutes.
3. Then remove pork, and put sugar. Sugar should be dissolved.
4. After that, place meat again into Instant Pot. Choose program Meat/Stew. Set a time at 1 hour.
5. Add soy sauce and red semi-dry wine. Stir it well.
6. Put onion, ginger, garlic, Sternanis, cinnamon into Instant Pot. Stir it well.
7. When the time is over, remove dish from Instant Pot.
8. Hun-shao-jo is ready. You can garnish it with vegetables or rice.

Nutrition:

- Calories: 165
- Fat: 14g
- Carbohydrates: 21g
- Protein: 11g

Beef ribs in balsamic honey marinade

Very tender and juicy dish. It is a crime not to try to cook it.

Prep time: 2 hours
Cooking time: 1 hour
Servings: 4

Ingredients:

- 2 pounds beef ribs
- 2 tablespoon balsamic
- 2 tablespoon honey
- 1 tablespoon Tabasco
- 2 tablespoon ketchup
- 1 tablespoon salt

Directions:

1. Wash and cut beef ribs
2. Put ribs in a bowl.
3. Add balsamic, tabasco, ketchup, honey and salt. Stir it well,
4. Leave it to marinate for about 2 hours.
5. Place meat into Instant Pot. Choose program Meat/Stew. Set a time at 1 hour
6. When the time is over, remove beef ribs from Instant Pot.
7. The dish is ready. You can garnish it with vegetables or rice.

Nutrition:

- Calories: 185
- Fat: 15g
- Carbohydrates: 23g
- Protein: 12g

Meat in Mexican style

Having tried this dish, it is impossible to remain indifferent. Definitely, you will ask for additional portion.

Prep time: 2 hours
Cooking time: 50 minutes
Servings: 2

Ingredients:

- 1 pound beef
- 8 ounce tomatoes
- 8 ounce zucchini
- 2 onions
- 2 eggplants
- 4 ounce tomato paste
- ¼ cup olive oil
- 1 chili pepper
- 1 clove of garlic
- 1 tablespoon salt

Directions:

1. Chop chili pepper and garlic.
2. Slice eggplants and season them with salt.
3. Peel tomatoes.
4. Dice all vegetables.
5. Dice beef.
6. Place beef and vegetables into Instant Pot. Choose program Meat/Stew. Set a time at 1 hour.
7. Remove the dish from Instant Pot. Let it cool down.
8. Meat in Mexican style is ready. Garnish it with vegetables of rice.

Nutrition:

- Calories: 210
- Fat: 12g
- Carbohydrates: 23g
- Protein: 10g

Rabbit in white wine and sour cream with mushrooms

The combination of garlic, greenery and white dry wine won't let you indifferent. Obviously, the rabbit is out of rivalry.

Prep time: 5 minutes
Cooking time: 1 hour
Servings: 4

Ingredients:

- 2 pound rabbit
- ¾ pound sour cream
- 1 tablespoon salt
- 1 tablespoon black pepper
- 4 cloves of garlic
- dill and parsley
- ¾ pound mushrooms (white)
- 3 tablespoon olive oil.
- 3 tablespoon butter
- 1 onion
- ½ cup white dry wine

Directions:

1. Cut rabbit into small pieces.
2. Season with salt and pepper.
3. Place rabbit into Instant Pot. Choose program Meat/Stew.
4. Set a time at 1 hour
5. Add sour cream to meat.
6. Add parsley and chopped garlic.
7. After 40 minutes put mushrooms above the meat.
8. Mix honey and wine.
9. Add to meat.
10. When the time is over, remove the dish from Instant Pot
11. The dish is ready. Serve it hot and with vegetables or rice.

Nutrition:

- Calories: 195
- Fat: 12g
- Carbohydrates: 19g
- Protein: 10g

Veal in coffee breadcrumbs

Veal and coffee is one of the most successful combinations. It may sound strange but you cannot say like that if you've never tried it!

Prep time: 20 minutes
Cooking time: 50 minutes
Servings: 2

Ingredients:

- 1 pound veal
- 4 tablespoon coffee
- 2 tablespoon pepper
- 1 tablespoon mustard
- 6 tablespoon sour cream.
- 1 tablespoon salt

Directions:

1. Cut veal into small pieces.
2. Smear each piece with sour cream and mustard.
3. Put meat into fridge for 20 minutes.
4. Mix coffee with salt and pepper.
5. Dredge veal with a mixture of coffee, salt and pepper.
6. Place meat into Instant Pot. Choose program Meat/Stew.
7. Set a time at 40 minutes.
8. Remove the meat when the time is over.
9. Veal in coffee breadcrumbs is ready. Bon appetite!

Nutrition:

- Calories: 205
- Fat: 14g
- Carbohydrates: 19g
- Protein: 10g

Pork with mustard and yoghurt

Sounds intriguing, doesn't' it? Let's check the tenderness of pork and sweetness of its sauce!

Prep time: 20 minutes
Cooking time: 50 minutes
Servings: 2

Ingredients:

- 2 pound pork
- 1 pound dried apricots
- 1 pound prunes
- 2 tablespoon yogurt
- 1 tablespoon mustard
- 1 tablespoon basil
- Rosemary
- 1 tablespoon salt

Directions:

1. Wash pork and make small holes.
2. Put dried apricots and prunes into these holes.
3. Season pork with salt.
4. Smear meat with yoghurt and mustard.
5. Sprinkle with basil.
6. Put it aside for 20 minutes
7. Place meat into Instant Pot. Choose program Meat/Stew.
8. Set a time at 40 minutes.
9. Remove the meat when the time is over.
10. Pork with mustard and yoghurt is ready. Bon appetite!

Nutrition:

- Calories: 210
- Fat: 13g
- Carbohydrates: 19g
- Protein: 11g

Pork in apple juice with pumpkin

These are appetizing chops. Thanks to apple juice, meat is very soft, juicy, with a light apple flavor

Prep time: 20 minutes
Cooking time: 1 hour
Servings: 2

Ingredients:

- 1 pound pork
- 2 cup apple juice
- 1 Bulgarian pepper
- 2 red onions
- 2 cloves of garlic
- ¾ pound pumpkin
- Spices
- 1 tablespoon salt
- 1 tablespoon black pepper

Directions:

1. Cut pork into pieces and tenderize each peace.
2. Drizzle meat with apple juice.
3. Slice pumpkin and Bulgarian pepper.
4. Chop garlic.
5. Place meat and vegetables into Instant Pot. Choose program Meat/Stew.
6. Set a time at 1 hour.
7. Remove the meat when the time is over.
8. Pork in apple juice with pumpkin is ready. Bon appetite!

Nutrition:

- Calories: 175
- Fat: 12g
- Carbohydrates: 16g
- Protein: 7g

Sweet pork ribs

Delicious and tender ribs are great variant for a winter meal!

Prep time: 1 hour
Cooking time: 1 hour
Servings: 4

Ingredients:

- 2 pound pork ribs
- 3 tablespoon soy sauce
- 2 tablespoon tomato paste
- 2 tablespoon honey.
- 2 tablespoon sugar.
- 1 tablespoon mustard
- 2 tablespoon vinegar.
- 3 cloves of garlic

Directions:

1. Mix soy sauce, tomato paste, honey and sugar. Make a marinade.
2. Wash and cut pork ribs.
3. Put pork into marinade and put it aside for 1 hour.
4. Smear meat with mustard
5. Place meat into Instant Pot. Choose program Meat/Stew.
6. Set a time at 1 hour.
7. Add vinegar.
8. Remove the meat when the time is over.
9. Sweet pork ribs are ready. Garnish it with vegetables.

Nutrition:

- Calories: 205
- Fat: 16g
- Carbohydrates: 26g
- Protein: 10g

Mutton in milk

As tender as never. Juicy mutton can surprise you.

Prep time: 1 hour
Cooking time: 1 hour
Servings: 4

Ingredients:

- 2 pounds mutton
- 2 carrots
- 1 onion
- 5 cloves of garlic
- Rosemary
- 2 cups Milk
- 2 tablespoon flour
- 2 tablespoon pepper
- 2 tablespoon salt
- 2 tablespoon adjika sauce

Directions:

1. Slice onion
2. Dice carrots
3. Cut mutton into pieces.
4. Place meat into Instant Pot. Choose program Meat/Stew.
5. Set a time at 1 hour.
6. Add vegetables.
7. Add milk.
8. After 10 minutes, add adjika sauce and flour. Stir it well.
9. Remove the meat when the time is over.
10. Mutton in milk is ready. Garnish it with vegetables.

Nutrition:

- Calories: 195
- Fat: 14g
- Carbohydrates: 24g
- Protein: 7g

Azu from beef

The recipe is very simple and even a beginner could implement it. Tender beef with…With what?

Prep time: 5 minutes
Cooking time: 1 hour
Servings: 4

Ingredients:

- 2 pounds beef
- 6 cucumbers
- 2 tablespoon tomato paste
- ¾ cup brine
- 5 cloves of garlic
- 2 tablespoon black pepper
- 2 tablespoon vegetable oil
- 1 onion

Directions:

1. Cut beef into pieces.
2. Season with pepper.
3. Peel and slice cucumbers.
4. Chop garlic and onion
5. Place beef and cucumbers, garlic and onion into Instant Pot. Choose program Meat/Stew.
6. Set a time at 1 hour.
7. Add tomato paste.
8. Add brine.
9. Remove the meat when the time is over.
10. Azu from beef is ready. Garnish it with vegetables and rice.

Nutrition:

- Calories: 165
- Fat: 13g
- Carbohydrates: 22g
- Protein: 9g

Mutton with mustard sauce

The recipe is very simple and even a beginner could implement it. Tender beef with…With what?

Prep time: 5 minutes
Cooking time: 1 hour
Servings: 4

Ingredients:

- 2 pound mutton
- 8 ounce breadcrumbs
- 4 ounce mustard
- ½ cup cream
- 4 ounce thyme
- 4 ounce oregano.
- 5 Potatoes
- 1 tablespoon vegetable oil
- 1 tablespoon salt

Directions:

1. Cut mutton.
2. Season with salt and pepper.
3. Dredge meat with breadcrumbs
4. Place meat into Instant Pot. Choose program Meat/Stew.
5. Set a time at 1 hour.
6. Meanwhile make a sauce.
7. Mix mustard and cream.
8. Add thyme and oregano. Stir it well.
9. Peel and cut potatoes.
10. After 40 minutes put potatoes and sauce into Instant Pot.
11. Remove the dish from Instant Pot when the time is over.
12. Mutton with mustard sauce is ready. Enjoy your meal!

Nutrition:

- Calories: 180
- Fat: 12g
- Carbohydrates: 21g
- Protein: 11g

Mutton with quince and vegetables

It is quite interesting combination of tastes. Check it by yourself.

Prep time: 5 minutes
Cooking time: 1 hour
Servings: 2

Ingredients:

- 1 pound mutton
- 8 ounce quince
- 4 ounce celery
- 2 onion
- 2 carrots
- 4 ounce thyme
- 2 tablespoon salt
- 2 tablespoon pepper
- ½ cup white dry wine

Directions:

1. Cut mutton.
2. Season it with salt and pepper.
3. Slice onion.
4. Add onion, thyme and celery to meat.
5. Slice quince and carrots
6. Place meat into Instant Pot. Choose program Meat/Stew.
7. Set a time at 1 hour.
8. After 40 minutes place the rest of ingredients into Instant Pot.
9. When the time is over, remove the dish from Instant Pot.
10. Mutton with quince and vegetables is ready. Sprinkle it with wine.

Nutrition:

- Calories: 230
- Fat: 14g
- Carbohydrates: 23g
- Protein: 11g

Sweet pork fillet

Not quite the usual version of pork fillet with dried fruits and bacon. It will be a delicious pork with an appetizing ruddy crust, delicious sauce and for a very short time!

Prep time: 12 hours
Cooking time: 1.5 hour
Servings: 4

Ingredients:

- 2 pound pork fillet
- 4 ounce dried fruits
- 4 ounce bacon
- 1 apple
- 2 celeries
- 1 onion
- 2 tablespoon cognac
- 4 ounce cranberry
- 4 ounce quince
- ½ cup water
- 2 tablespoon soy sauce

Directions:

1. Dice dried fruits and apple.
2. Dice bacon.
3. Mix bacon, apple, dried fruits, cranberries and grind everything.
4. Add cognac. Stir it well.
5. Put this mass into fridge for 12 hours.
6. Wash pork fillet. Tenderize it slightly.
7. Remove mass from the fridge and put it onto the pork fillet.
8. Seal the edges.
9. Place meat into Instant Pot. Choose program Meat/Stew.
10. Set a time at 1.5 hour.
11. When the time is over, remove the dish from Instant Pot.
12. Sweet pork fillet is ready. Sprinkle it with soy sauce. Garnish with rice.

Nutrition:

- Calories: 245
- Fat: 14g
- Carbohydrates: 26g
- Protein: 10g

Pork in sesame

The ideal combination of everything – meat, sesame, spices. The most ideal in the whole world.

Prep time: 12 hours
Cooking time: 1.5 hour
Servings: 4

Ingredients:

- 2 pound pork
- 5 tablespoon sesame
- 4 tablespoon honey
- 5 tablespoon Teriyaki sauce
- 1/3 cup water
- 2 tablespoon pepper
- 1 onion
- 1 clove of garlic

Directions:

1. Mix honey and Teriyaki sauce in a bowl.
2. Add chopped garlic
3. Cut pork.
4. Season it with pepper.
5. Slice onion.
6. Dip pork into mixture of honey and Teriyaki sauce.
7. Dredge meat piece with sesame.
8. Place meat into Instant Pot. Choose program Meat/Stew.
9. Set a time at 1 hour.
10. Place onion above the meat.
11. When the time is over, remove the dish from Instant Pot.
12. Let it cool down.
13. Pork in sesame is ready. You can serve it with rice or vegetables.

Nutrition:

- Calories: 265
- Fat: 16g
- Carbohydrates: 28g
- Protein: 11g

FREE BONUS GIFT

Dear friend!
Thank you so much for buying my book and supporting my next cookbooks which I hope you will enjoy as well. In order to thank you I am very happy to present you a gift cookbook: "General Cooking: 220 Best Recipes".

Please follow this link to get instant access and DOWNLOAD your BONUS:
https://goo.gl/hZieJi

(No subscription or other additional actions required)

Made in the USA
Lexington, KY
04 November 2017